AF678523

Basic Engineering Circuit Analysis

Basic Engineering Circuit Analysis

Editor

Tudor Volkov

scitus academics

Basic Engineering Circuit Analysis

Edited by **Tudor Volkov**

Printed in 2017

ISBN: 978-1-68117-111-1
Library of Congress Control Number: 2015954090

© 2016 by
SCITUS Academics LLC,
616, Corporate Way, Suite 2, 4766,
Valley Cottage, NY 10989

www.scitusacademics.com

This book contains information obtained from highly regarded resources. Copyright for individual articles remains with the authors as indicated. All chapters are distributed under the terms of the Creative Commons Attribution License, which permits unrestricted use, distribution, and reproduction in any medium, provided the original author and source are credited.

Notice

Reasonable efforts have been made to publish reliable data and views articulated in the chapters are those of the individual contributors, and not necessarily those of the editors or publishers. Editors or publishers are not responsible for the accuracy of the information in the published chapters or consequences of their use. The publisher believes no responsibility for any damage or grievance to the persons or property arising out of the use of any materials, instructions, methods or thoughts in the book. The editors and the publisher have attempted to trace the copyright holders of all material reproduced in this publication and apologize to copyright holders if permission has not been obtained. If any copyright holder has not been acknowledged, please write to us so we may rectify.

Preface

Basic engineering circuit analysis is a process through which engineers are able to set up electrical models and are able to express physical situations in terms of mathematical relations.It would be absolutely absurd and a dangerous idea for anybody thinking of going for a deep sea exploration without first learning how to swim. The same is true in the field of electrical engineering and electronics. Without a thorough knowledge of the basics, success in the relevant fields may be quite farfetched. In an electrical circuit the process of studying and analyzing the various electrical quantities involved, especially the nodal voltages and currents through calculations, is known as circuit analysis.

In this book we will go through a detailed study of a few circuit configurations and try to solve the problems involved in these elementary electrical circuits through illustrative examples. Circuit analysis is the fundamental gateway for computer and electrical engineering majors. Basic Engineering Circuit Analysis has long been regarded as the most dependable book.Circuit analysis remains the starting point for many future engineers who wish to work in this field.

Table of Contents

Chapter 1

Fractional RC and LC Electrical Circuits *

Gómez-Aguilar José Francisco[1,] *, Rosales-García Juan*[3,] *, Guía-Calderón Manuel*[4,]

Departamento de Ingeniería Eléctrica División de Ingenierías Campus Irapuato-Salamanca Universidad de Guanajuato
Razo-Hernández José Roberto[2,]
Departamento de Electromecánica Instituto Tecnológico Superior de Irapuato

ABSTRACT

In this paper we propose a fractional differential equation for the electrical *RC* and LC circuit in terms of the fractional time derivatives of the Caputo type. The order of the derivative being considered is $0 < \gamma < 1$. To keep the dimensionality of the physical parameters *R, L, C* the new parameter a is introduced. This parameter characterizes the existence of fractional structures in the system. A relation between the fractional order time derivative γ and the new parameter σ is found. The numeric Laplace transform method was used for the simulation of the equations results. The results show that the fractional differential equations generalize the behavior of the charge, voltage and current depending of the values of γ. The classical cases are recovered by taking the limit when $\gamma = 1$. An analysis in the frequency domain of an *RC* circuit shows the application and use of fractional order differential equations.

INTRODUCTION

Although the mathematical foundation of Fractional Calculus (FC) was established over 200 years ago, there remains a subject quite new to mathematicians. FC, involving derivatives and integrals of non-integer order, is the natural generalization of the classical calculus (Oldham and Spanier, 1974; Miller and Ross, 1993; Samko *et al*, 1993; Podlubny *et al*, 1999; Uchaikin, 2013). Many physical phenomena have "intrinsic" fractional order descriptions and so FC is necessary in order to explain them. In many applications FC provides more accurate models of the physical systems than ordinary calculus do. Since its success describing anomalous diffusion (Wyss *et al*, 1986; Hilfer, 2000), (Meteler and Klafter, 2000; Agrawal *et al.*,2004) non-integer order calculus both in one- and multi-dimensional spaces, it has become an important tool in many areas of physics, mechanics, chemistry, engineering, finances electromagnetism and bioengineering (Petras, 2010; Obeidat *et al*, 2011; Hilfer, 1974; West *et al*, 2003; Rosales*et al*, 2012; Magin *et al.*, 2006; Gómez *et al.*, 2012c; Gómez *et al.*, 2012d; Caputo and Mainardi, 1971). Fundamental physical considerations in favor of the use of models based on derivatives of non-integer order are given in Westerlund (1994), Veliev (2004)and Baleanu *et al.* (2010). The advantage of using fractional order systems compared with systems of integer order is that the former has infinite memory, while others have finite memory. This is the main advantage of FC in comparison with the classical integer-order models, in which such effects are in fact neglected.

To analyze the dynamical behavior of a fractional system it is necessary to use an appropriate definition of the fractional derivative. In fact, the definitions of the fractional order derivative are not unique and there exist several definitions, including Grü;nwald-Letnikov, Riemann-Liouville, Weyl, Riesz, and the Caputo representation. In the Caputo case, the derivative of a constant is zero and we can properly define the initial conditions for the fractional differential equations so that they can be handled analogously to the classical integer case. Caputo derivative implies a memory effect by means of a convolution between the integer order derivative and a power of time (Podlubny, 1994;Gutiérrez *et al.*, 2010). For these reasons, in this paper we prefer to use the Caputo fractional derivative defined by:

$$D^{\gamma} f(t) = \frac{1}{\Gamma(n-\gamma)} \int_{0}^{t} \frac{f^{(n)}(\tau)}{(t-\tau)^{\gamma+1-n}} d\tau, \qquad (1)$$

where n-1 < γ < *n* (*n* is N), and $f^{(n)}(x)$ represents the derivative of order *n*, real function evaluated in *t*. Working with this definition is important because of the ability to be implemented numerically (Podlubny, 1994). The Caputo derivative has the following property: if *f* (t) is a constant, then its derivative is zero. This does not happen with other representations. Another very important feature in the form of Caputo fractional derivative is that its Laplace transform is:

$$L\{D^{\gamma} f(t)\} = s^{\gamma} F(s) - \sum_{k=0}^{n-1} f^{(k)}(0)\, s^{\gamma-k-1} \tag{2}$$

From equation (2), we can see that the representation of the Caputo derivative in Laplace domain using the initial conditions $f^{(k)}(0)$ where *k* is integer. If the initial conditions are zero, this reduces to:

$$L\{D\gamma f(t)\}=s\gamma F(s) \tag{3}$$

This is consistent with the usual definition of the Laplace transform when γ is an integer. In general, a fractional order differential equation has the form

$$\sum_{k=0}^{n} a_k D^{\gamma k} f(t) = g(t) \tag{4}$$

where $\gamma_k > \gamma_{k-1}$ and *ak* are any real numbers, g(t) can be seen as the source of a dynamic system. Inverse transform 0 < γ < 1 requires the introduction of a special function. The Mittag-Leffler function, where Γ is the gamma function, which is defined as

$$E_{a,b}(t) = \sum_{m=0}^{\infty} \frac{t^{m}}{\Gamma(am+b)}, (a, b > 0), \tag{5}$$

from (5) if *a* = 1, *b* = 1, we obtain the expression $E_{1,1}$ (t) = *e.t* The Mittag-Leffler function is a generalization of the exponential function.

NUMERICAL LAPLACE TRANSFORM

The Laplace transform is a useful tool for analyzing linear systems because it simplifies the problem of dealing with differential equations in the time domain by converting them into algebraic equations within the frequency domain. The numerical Laplace transform (NLT) is essentially a modified discrete Fourier transform (DFT) through a windowing function (Gibbs phenomenon) and a stability factor (aliasing), (Proakis and Manolakis, 1996). Development of the NLT and its application to the analysis of systems has been well documented over the past 40 years (Ramirez *et al.*, 2004; Wilcox and Gibson, 1998). However, its use has traditionally been limited to the analysis of problems where the result can be expressed in terms of simple functions that allow the use of tables of Laplace transforms (Moreno and Ramírez, 2008). When using discrete techniques in the frequency domain, computation time becomes an important factor since it requires a certain amount of time to transform the data from the frequency domain to the time domain or vice versa. However, by using the fast Fourier transform (FFT) the time necessary for computation is greatly reduced and as a result the techniques of analysis in the frequency domain become an attractive option.

Sheng investigated the validity of applying numerical inverse Laplace transform algorithms in fractional calculus (Sheng y Chen, 2011). In a paper (2012b), an overview of a methodology based on the NLT and applied to the analysis of electromagnetic transient phenomena in power systems described by differential equations of fractional order, a Newton-type methodology to calculate either the transient or the periodic steady state is used and the definition of Caputo fractional derivative is applied. To electrical networks including nonlinear reactors and electronic devices.

In the present work we are interested in the study of a simple electrical circuit consisting of a resistor, an inductor and a capacitor, in the framework of the fractional derivative applying the method of the NLT for the simulations.

Description

Using the Kirchhoff voltage law and the circuit of Figure 1, we have:

$$E(t)=V_e(t)+V_C(t), \quad (6)$$

where E(t) is the source voltage, $V_e(t)$ is the voltage in the fractal element and $V_C(t)$ is the voltage in the capacitor. The voltage in the fractal element is

$$V_e(t) = \frac{e}{\sigma_e^{(1-\gamma)\beta}} \frac{d^{\beta\gamma} q(t)}{dt^{\beta\gamma}}, \tag{7}$$

where β {1, 2} is a parameter that determines whether the element *e* is a resistor or inductor, $0 < \gamma \leq 1$, the product $\beta_\gamma$determines the order of the fractional differential equation *e* {*R*, *L*} defines the characteristics of e, a is a parameter that determines the fractional structures of *e*.

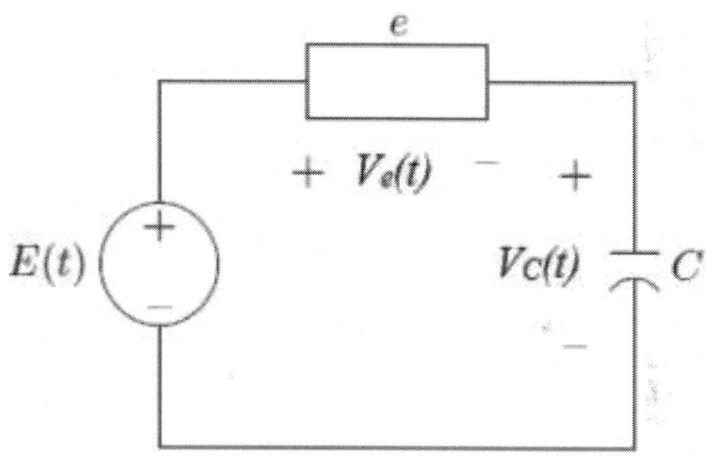

Figure 1. Circuit with a fractal element and a capacitor

Fractional electrical circuit RC

Taking in (7)*e* = *R*, β = 1, σ*e* = σ*R*, the fractional differential equation for *RC* circuit is

$$\frac{R}{\sigma_R^{1-\gamma}} \frac{d^{\gamma} q(t)}{dt^{\gamma}} + \frac{q(t)}{C} = E(t), (0 < \gamma \leq 1) \tag{8}$$

Recently Gómez *et al.* (2012a) proposed a systematic way to construct fractional differential equations for the physical systems. Such a way consists in analyzing the dimensionality of the ordinary derivative operator and trying to bring it to a fractional derivative operator consistently.

The introduction of the parameter σR allows dimensional consistency when $[\sigma R] = sec$

$$\left[\frac{1}{\sigma_R^{1-\gamma}}\frac{d^\gamma}{dt^\gamma}\right] = \frac{1}{sec^\gamma}, (0 < \gamma \leq 1) \tag{9}$$

with $\gamma = 1$ the equation (9) becomes an ordinary differential equation the first order with respect to time t This is only true, as stated above, if σ_R has dimensions of seconds.

Using the expression (8), the fractional differential equation for a circuit of a capacitor and a resistor has the form

$$\frac{d^\gamma q(t)}{dt^\gamma} + \frac{q(t)}{\tau_\gamma} = \frac{C}{\tau_\gamma}E(t), \tag{10}$$

with

$$\tau_\gamma = \frac{RC}{\sigma_R^{1-\gamma}} \tag{11}$$

The constant τ_γ also can be called fractional time constant due to its dimensionality $[sec]^\gamma$. When $\gamma = 1$, equation (11) recovers the ordinary time constant, ie., $\tau_1 = \tau = RC$.

The γ parameter, which represents the order of the fractional time derivative in (10), can be related to the parameter σ_R, which characterizes the presence of fractional structures (fluctuations) in the system. In this case the empiric relationship is given by the expression

$$\gamma = \frac{\sigma_R}{RC}, \tag{12}$$

thus, the magnitude

$$\delta = 1 - \gamma, \tag{13}$$

characterizes the existence of fractional structures in the system.

It can be seen this way: if γ = 1, then, from (12) we have $a_R = \tau = RC$ and hence 5 = 0 in(13), which means that the system does not have fractional structures, it is an ordinary*RC* circuit. However, in the range 0 < γ < 1, or 0 < σ < τ, the quantity 5 increases and tends to unity, since in the system each is again a fractional structure.

Assuming zero initial conditions, *q* (0) = 0 and *E* (t) = sin *ωt*, the Laplace transform applied to (10) results in

$$Q(s) = \frac{C}{\tau_\gamma\left(s^\gamma + \frac{1}{\tau_\gamma}\right)}(E(s)) \tag{14}$$

The voltage in the capacitor is

$$V_C(s) = \frac{1}{\tau_\gamma\left(s^\gamma + \frac{1}{\tau_\gamma}\right)}(E(s)) \tag{15}$$

The current can be obtained applying inverse numeric Laplace transform in (14) and differentiating with respect to time.

Fractional electrical circuit LC

$$\frac{L}{\sigma_L^{2(1-\gamma)}}\frac{d^{2\gamma}q(t)}{dt^{2\gamma}} + \frac{q(t)}{C} = E(t).(0 < \gamma \leq 1) \tag{16}$$

Using the expression (16), the fractional differential equation for a circuit containing a capacitor and an inductor has the form

$$\frac{d^{2\gamma}q(t)}{dt^{2\gamma}} + \frac{q(t)}{\tau_{L\gamma}} = \frac{C}{\tau_{L\gamma}}E(t) \tag{17}$$

Where

$$\tau_{L\gamma} = \frac{\sqrt{LC}}{\sigma_L^{1-\gamma}} \tag{18}$$

The constant τL_γ also can be called fractional time constant due to its dimensionality [*sec*]. When $\gamma = 1$, from (18) recovers the ordinary time constant, this is, $\tau_{L1} = \tau_L = \sqrt{LC}$.

The γ parameter, as before, that represents the order of fractional time derivative in (17), can be related to the parameterσL, which characterizes the presence of fractional structures (fluctuations) in the system. In this case the empiric relationship is given by the expression

$$\gamma = \frac{\sigma_L}{\sqrt{LC}}, \qquad (19)$$

thus, the magnitude

$$\delta = 1 - \gamma, \qquad (20)$$

characterize the existence of fractional structures in the system.

You can see it this way: if $\gamma = 1$, then, from (19) we have $\sigma_L^2 = LC$ and therefore $\delta = 0$ in(20), which means that the system has no fractional structures, is an ordinary *LC* circuit. However, in the range $0 < \sigma < \sqrt{LC}$, (undamped natural frequency), the magnitude δ increases and tends to the unit, as in the system are increasingly found fractional structures.

Assuming the initial conditions $q(0) = q_0$, $q(0) = 0$ and power supply $E(t) = \sin \omega(t)$, the Laplace transform applied to (17) results,

$$Q_k(s) = \frac{CE(s) + q_0 \tau_{L\gamma} s}{\tau_{L\gamma}\left(s^{2\gamma} + \frac{1}{\tau_{L\gamma}}\right)}. \qquad (21)$$

The voltage in the capacitor is

$$V_C(s) = \frac{CE(s) + q_0 \tau_{I\gamma} s}{C\tau_{I\gamma}\left(s^{2\gamma} + \frac{1}{\tau_{I\gamma}}\right)}. \tag{22}$$

The current can be obtained applying inverse numeric Laplace transform in (21) and differentiating with respect to time.

RESULTS

All simulations used MATLAB 7.5. (R2007b), Figure 2, Figure 3 and Figure 4 show the simulation for *RC* circuit γ = 1, γ = 0.98, γ = 0.96, for the charge, voltage and current, respectively, using *R* = 1 *pu*, *C* = 0.1 *pu* and ω = 2π60.

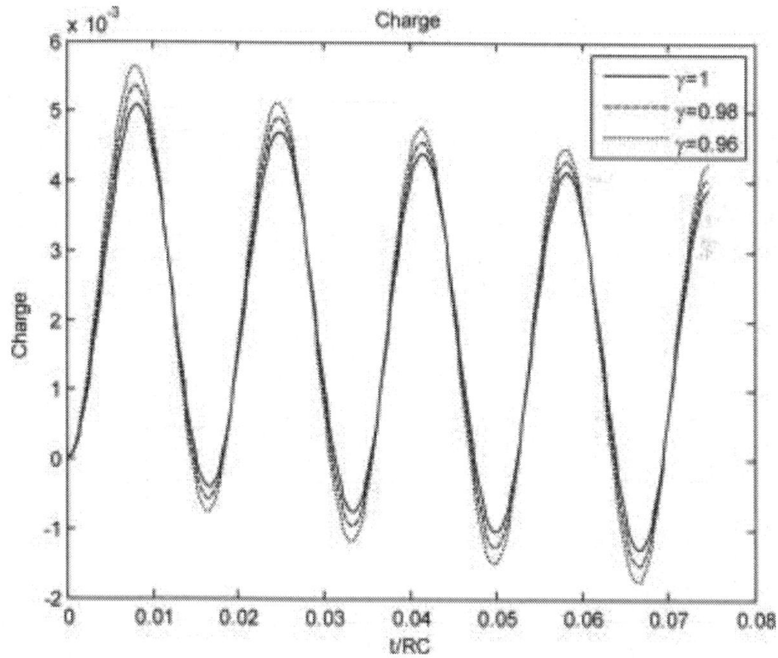

Figure 2. Graph of the charge on the capacitor in the *RC* fractional circuit for different values of γ

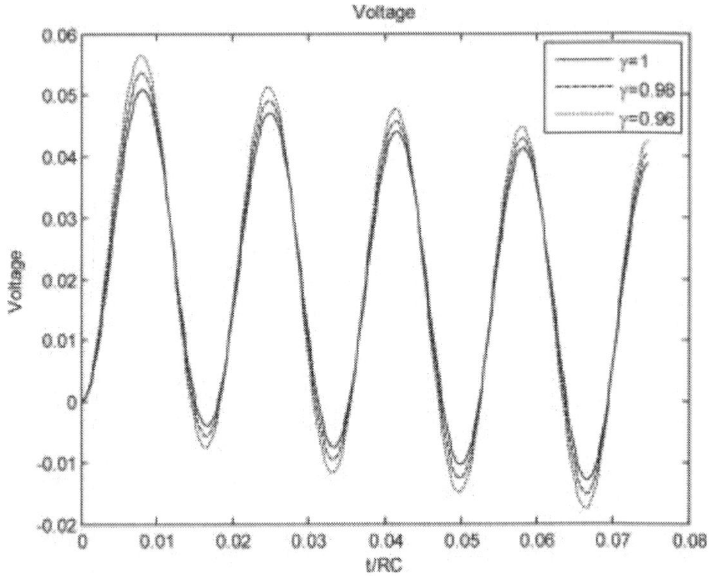

Figure 3. Graph of the voltage on the capacitor in the *RC* fractional circuit for different values of γ

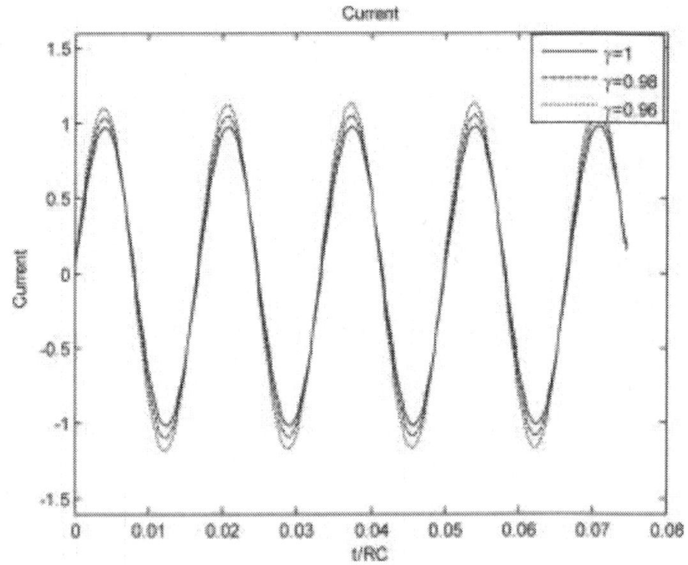

Figure 4. Graph of the current in the *RC* fractional circuit for different values of γ

Figure 5 show the simulation for LC circuit $\gamma = 1$, $\gamma = 0.98$, $\gamma = 0.96$, for the current, using L = 0.001 pu, C = 0.1 pu and $\omega = 2\pi 60$.

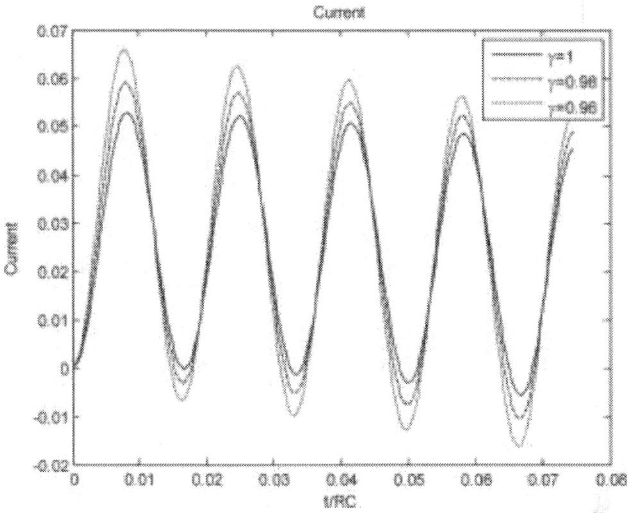

Figure 5. Graph of the current in the *LC* fractional circuit for different values of γ

Spectroscopy applied to a RC circuit

The electrical impedance spectroscopy technique applies a potential difference between the two electrodes by passing a low power alternating current through the sample and this is compared with the voltage and current detected to the output. To obtain the electrical impedance spectra a Solartron® 1260 was used (Figure 6). The fidelity of the frequency sweep for these tests was important since it shows the characteristic spectrum of the sample, which is necessary for comparing with the electrical parameters of an equivalent circuit. The frequency range used was from 10 Hz to 100 kHz. Once the measurements were represented in a Nyquist diagram, we obtained representations of equivalent electrical circuits via the software ZView®. The method of "instant adjustment" was used to fit the data to predefined circuit models. A voltage of 25 mV was applied across the *RC* circuit.

Figure 6. Solartron® 1260

To determine the equivalent equation, the impedance is determined, which is given by the following formula in the complex frequency domain.

$$Z(s) = \frac{V(s)}{I(s)} \quad (23)$$

Applying Kirchhoff laws to the circuit of Figure 1, we have:

$$v = R_s i + Vc_p \quad (24)$$

$$i = i_R + i_c \quad (25)$$

Before applying the Laplace transform of (24) and (25) must make the following considerations

$$i_R = \frac{Vc_p}{R_p} \quad (26)$$

$$i_c = C_p \frac{dV_c}{dt} \quad (27)$$

In applying the definition of fractional derivative (Podlubny, 1994), we have:

$$i_R = \frac{Vc_p}{R_p} \tag{28}$$

$$i_c = C_p \frac{d^{\gamma} Vc_p}{dt^{\gamma}} \tag{29}$$

Substituting (28) and (29) into (25), we obtain:

$$v = R_s i + Vc_p \tag{30}$$

$$i = \frac{Vc_p}{R_p} + C_p \frac{d^{\gamma} Vc_p}{dt^{\gamma}} \tag{31}$$

Applying Laplace transform to (30) and (31) is obtained:

$$V(s) = R_s I(s) + Vc_p(s) \tag{32}$$

$$I(s) = \frac{Vc_p(s)}{R_p} + \frac{C_p}{\sigma^{1-\gamma}} s^{\gamma} Vc_p(s) \tag{33}$$

Finally from (32) and (33) we have:

$$Z(s^{\gamma}) = R_s + \frac{R_p}{1 + \frac{R_p C_p}{\sigma^{1-\gamma}} s^{\gamma}} \tag{34}$$

If, s = (*jw*), we have:

$$Z(j\omega)^{\gamma} = R_s + \frac{R_p}{1 + \frac{R_p C_p}{\sigma^{1-\gamma}} (j\omega)^{\gamma}}. \tag{35}$$

The equation (35) is the result of the fractional temporal operator in the equation for the*RC* equivalent circuit, this general formula includes an arbitrary constant sigma and in the case particular to take the value of σ = *RPCP* is reduced to the Cole model (1941) andGómez *et al.* (2012). On the other hand, if we make γ = 1, we obtain the *RC* circuit ideal.

Figure 7 show the comparison of the Nyquist diagram resulting between integer order equation (23), the spectroscopy and fractional equation (35), fractional exponent γ = 0.9825. This value was obtained by a least-squares fit.

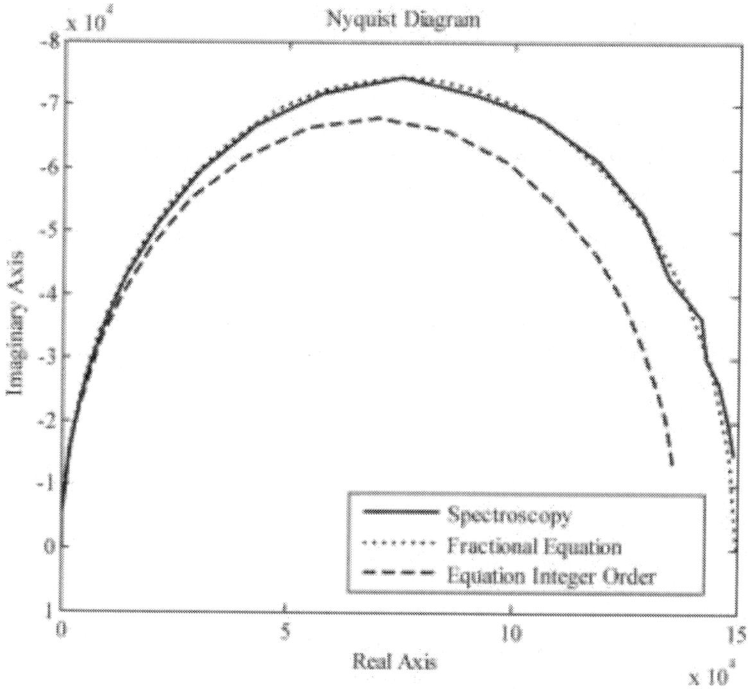

Figure 7. Nyquist diagram, comparison between integer order equation (23), the spectroscopy and fractionalequation (35), fractional exponent γ = 0.9825

From the Figure 7, it can be seen that the fractional differential equation (35) obtained describes best the measurement of electrical impedance spectroscopy.

DISCUSSION

The Figures 2 and 4 show the charge and current in the *RC* circuit, respectively. FromFigure 3 can observe that as γ increases from 0.96 to 1, the fractal element behaves as a resistor of resistance *R* for $\gamma = 1$, intermediate values of γ determines behavior between a capacitor and a resistor. In the Figure 5 is observed a behavior inductive for $\gamma = 1$, intermediate values determine behavior between inductor and resistor.

It has been found that for every $0 < \gamma \leq 1$, the load stored in the *RC* circuit is directly proportional to the voltage on the capacitor; it follows that the behavior of the charge on the capacitor will give identical curves to those shown in Figure 3.

This current behavior can be induced from voltage in the capacitor, for $\gamma = 1$ will have a current value equal to the ratio between the supply voltage and resistance for $t = 0$, and will decrease with time until it reaches zero, a full-order RC circuit.

As to the fractional *LC* circuit, is observed in Figure 5, the current increases as the γ values decrease from 1 to 0.96 indicating losses in the *LC* series circuit, this can be seen because there is a damping voltage curves, typical of an *RLC* circuit with losses.

CONCLUSIONS

The fractional differential equation for the RC and LC circuits has been proposed. The relevant aspect of this work is the way to introduce the fractional derivatives operators, providing a systematic way to construct the fractional differential equations of any physical system keeping the dimensionality of the physical parameters. The order of the derivative being considered is $0 < \gamma \leq 1$. To keep the dimensionality of the physical parameters R, L, C the new parameter σ is introduced. This parameter characterizes the existence of fractal structures in the system (components that show an intermediate behavior between conservative and dissipative systems), the fractional time components change the time constant and affect the transient response of the system (Guía *et al.*, 2013). A relation between the fractional order time derivative γ and the new parameter σ is found. The numeric Laplace transform method was used for the simulation of the equations results. The classical cases are recovered by taking the limit when $\gamma = 1$. Also, the concept of fractional time constant has been introduced.

From the description of the fractional differential equation models can be noted that the representation of the Cole model is generated to solve the *RC* circuit in the formalism of fractional calculus. The simulations obtained from the fractional representation showing a better description than those obtained by the equations of integer order. The exponent of the fractional differential equation that best fits the *RC* circuit is $\gamma = 0.9825$.

With the approach presented here, it will be possible to have a better study of the transient effects in the electrical systems. Also, the electrical circuit will provide a robust framework for studying the bioelectrical response to transient stimuli, when an equivalent electrical circuit has been used.

ACKNOWLEDGEMENTS

This research was supported by CONACYT.

REFERENCES

1. Agrawal et al., 2004, O.P. Agrawal, J.A. Tenreiro-Machado, I. Sabatier, Fractional Derivatives and Their Applications, Nonlinear Dynamics, Springer-Verlag, Berlin (2004).
2. Baleanu et al., 2010, D. Baleanu, Z.B. Günvenc, J.A. Tenreiro-Machado, New Trends in Nanotechnology and Fractional Calculus ApplicationsSpringer (2010)
3. Caputo M., Mainardl F. Pure and Applied Geophysics, volume 91, 1971:134.147.
4. Cole ICS and Cole RAI Dispersion and Absorption in Ontario.: Alternating Current Characteristics. 1. Clem. Php. volume 9, 1941:341-351.
5. Gomez-Aguilar J.F. Rosak•-Carcia I.I. Bernal•Alvarado f.J., Ceadovaraga T. Guzmin.Cabrera ft. Fractional Mechuti-cal Oscillators. Atria/ Makana de Fiske, volume SS, 2012a: 524-537.
6. Comex I.F. Rosales J I. Banal IA. Cul.% M. Martinez I. Electro-magnetic Transient Analysis of Networks Described by Equa-tions of Fractional Order. Prespaceriat feu nu/. volume 3 (issue 6), 2012b:524-537
7. Comezaguibr F. RosaleiCarcia 1, Cula.C.0dertin Al, Bernal-Alvared0 I. Analysis of Equivalent Circuits for Cells: A Frac-tional Calculus Approach. Amiga illgaritrilt, linwrigucine y Tecnologia, UN MI. volume MIL 3012c: 375-384.
8. Gomez F. Bernal I Routes I. Cardm-a T. Modeling and Sunni.). Lion of Equivalent Cl(=lt% in Dewnption of Biolopwal Sys. tems. A Fractional Caktlllt& Approach. /puma: of Electrical Biointedanct, volume 3 (Issues 2-11), 2012d.

9. Guia M. Games F., Rosaka I. Ana/ysis on the Time and Prequel). cy Domain for the RC Electric Circuit of Fractional Order. Cent. Eur.1 Ploys., Springer, 2013.
10. Gutierrez RS, Rodin I.M. Tenroro•Machado J.T. Fractional Or-der Caladus: Basic Concepts and Fagiawriug Aplications, Itin-dawi Publishing Cosporatinn, Mathematical Problems in Engineering, 2010
11. Hilfer R. Appticeions of Fractional Calndus L Physics, Singapore, Ed. World Scientific. 2000.
12. Hilfer R. Phys. OW M. B. volume 101, 1974: 39144917.
13. Maps R.L Fractiosol Calcutta, in Sieenginoning. Begell Nome Pu-blisher, Rodding, 2006.
14. Matder R, Klafter I. Phr. Rep. volume 339, 2100:1-77.
15. Miller ICS. Ross B. An Introduction to the Fractional Calculus and Fractional Difh initial Equations, Wiky, New York, 1993
16. Moreno P and Ramirez A. Implementation of the Numerical La. place Transform a ReVie•. IEEE Irmo. Power Delivery, volu-me 23 (issue 4), 2008: 2599-2609.
17. Oldham KB. Spaniel j. 71w frocrioud Undid, Academic Press, New York, 1974. Cbeidat A., Gharibeh M. Mali M. Roman A. Evolution of a Cu. cram in a Resistor. Art. Cali'. App. Anal., Springer 211 1, pp. 2-14.
18. Petra I. Frachonal.Order Atenwaster•BasS Chuffs Circuit. MCC Transactions on Circuits end *Utopia: txprrus ariefr. wturne 57, 2010l-12
19. Podlubny I. fractional Deem:tail Lourtiont,, San Diego, Academic Proxy 1994.
20. Proakii J.C. and kLumlakii D.C. Digital Signal Promising. in Prim Algona:no and Appian/ran, 3rd ed. Upper Saddle R. ter, NJ. Prennce.11.01, 1996.
21. Ramirez A. C.Fmez P. Moreno P. Gutierrez A. Frequency Do-man Analysis of Elettromapwir Tranues Through the Numencal Laplace Transform on: The IEEE Cameral Meting, Denier, CO, 2004
22. Rocks IA. Cuia M. Comes J.F., Mach V.I. Eras-bona/ Electra-intimate Wave. Discontinuity. Nadincerint and Complexity, volume 14, 2012: 315-330
23. Samko SG. Kiltas A.A., Mantchey O.I. rr.nth i !wroth nil Or fins. Theory and Applinvian, Cordon and Breach Schnee Publishers, Langhorne, PA, 1993.
24. Sheng H. la Y., Olen Y.Q. Appbcabon of Numerical Inverse La-place Taredonn Alsonthms in Fractional Calculus. 1.frantlia hem, volume XIS (roue 2), 2011:315-330.
25. Uchaikin V. frothing! Daimlers fir Physician and Engineers, Springer 2013
26. Volley E.1., Engheta N. Fractional Cud Operator m Reflettion Pro-blenv. on 10th InL Conf. on Mathematical Methods in Den tromapttic Theory, Ukraine, September X01, pp. 14-17.
27. West &J., Bobgna M., Crigoliai P. Apia of frealatal Opinion, Berlin, Springer-Verbg, 2003
28. Weiterlund S Cause y, Report no. 940174 University of Kalmar, 1994

29. Wilcox D.I. Gibson IS Numerical Laplace Transformation and Inversion in the Malysas of Physical System.. hif. I. Nunn. Media, [mg- volume20, 19M:1507.1519.
30. Wyss W.J. Aga& Mir, volume 27, 19716:2782-2785.

CITATION

Gómez-Aguilar José Francisco, Rosales-García Juan, Guía-Calderón Manuel, Razo-Hernández José Roberto, Fractional RC and LC Electrical Circuits, Ingeniería, Investigación y Tecnología, Volume 15, Issue 2, April–June 2014, Pages 311-319, ISSN 1405-7743, http://dx.doi.org/10.1016/S1405-7743(14)72219-X.

Chapter 2

High Figure-of-Merit SOI Power LDMOS for Power Integrated Circuits

Yashvir Singh, Rahul Singh Rawat*

Department of Electronics & Communication Engineering, G. B. Pant Engineering College, Pauri Garhwal, Uttarakhand 246 194, India

ABSTRACT

The structural modifications in the conventional power laterally diffused metal-oxide-semiconductor field-effect transistor (LDMOS) are carried out to improve the breakdown voltage, on-resistance, gate-charge and figure-of-merits of the device with reduced cell pitch. The modified device has planer structure implemented on silicon-on-insulator which is suitable for low to medium voltage power integrated circuits. The proposed LDMOS consists of two gate electrodes placed vertically in two separate trenches build in the drift region and single source and drain contacts are taken on the top. The trench structure reduces the electric field inside the drift region and allow increased drift layer doping concentration leading to higher breakdown voltage, lower specific on-resistance, reduced gate-drain charge, and substantial improvement in the figure-of-merits. Using two-dimensional simulations, the performance of the proposed LDMOS is optimized and results are compared with the conventional LDMOS. Our simulation results show that the proposed device exhibits 110% higher breakdown voltage, 40% reduction in cell pitch, 19% lower specific on-resistance, 30% lower gate-to-drain charge leading to 5.5 times improvement in Baliga's figure-of-merit and 43% reduction in dynamic figure-of-merit over the conventional device.

INTRODUCTION

The power integrated circuits (PICs) have gained importance due to their advantages of small size, low cost, less power consumption, and increased reliability. The PICs operating in low to medium voltage range (60–100 V) are used in many applications [1],[2], [3] and [4] such as automotive electronics, light-emitting diode drivers, plasma display panels, PC peripheral, and portable power management products. For fabricating PICs, silicon-on-insulator (SOI) technology is a preferred choice because it provides superior isolation between various devices on the chip. In such circuits, the key power device is laterally diffused metal-oxide-semiconductor field-effect transistor (LDMOS). A power LDMOS is designed to achieve higher breakdown voltage while minimizing specific on-resistance in order to reduce conduction losses. The specific on-resistance (R_{ON}) is defined as the product of total on-resistance and area of the device. Both the breakdown voltage (V_{BR}) and the on-resistance of SOI LDMOS are dependent on the length and doping of the drift region. A long drift region length and low doping is required to achieve high breakdown voltage in a conventional LDMOS (CLDMOS) which unfortunately, increases the on-resistance of the device. Conversely, a shorter drift region length with higher doping reduces the on-resistance but adversely affects the breakdown voltage. Therefore, there is always a trade-off between the breakdown voltage and the on-resistance. One measure of this trade-off is the Baliga's figure-of-merit (BFOM = V_{BR}^2/R_{ON}) which should be as high as possible for a better designed LDMOS. For PICs applications, it is also important to estimate the switching performance of an LDMOS. In order to reduce the switching losses, the gate-charge in general and the gate-to-drain charge (Q_{gd}) in particular should be minimized for a given on-resistance. The switching performance of an LDMOS is evaluated by dynamic figure-of-merit (FOM = $R_{ON}.Q_{gd}$). For improving the static and dynamic performance of a CLDMOS, several modifications in the conventional design have been reported [5], [6], [7],[8] and [9] by incorporating the trench structures in the drift region. These devices have successfully demonstrated to achieve low on-resistance and high breakdown voltage by increasing the drift region doping while reducing the electric field inside the drift region. Furthermore, a lower cell pitch of the device is desirable to get cost advantage with increased packaging density of PICs.

A lateral trench gate LDMOS structure was reported [8] which utilized a trench for channel formation in the drift layer on SOI and achieved a BFOM of 6.68 MW/cm^2 with a cell length of 6 μm. Another trench-gate LDMOS was proposed [9] on thin SOI consisting of two trenches in the drift region on both sides of the p-body region. This device is demonstrated to have a BFOM of 12.87 MW/cm^2 for a cell pitch of 5 μm and also provides significant improvement in switching performance. In this structure, further reduction in the cell pitch of the LDMOS was limited by the lateral dimensions of the device due to the dual drain contacts symmetrically placed on both sides of the trench-structure. In order to improve the BFOM with reduced cell pitch, we propose a trench-gate LDMOS structure called TGLDMOS in which two trenches are created in the drift region on both side of p-body where gate electrodes are placed. The unit cell of the device contains only single drain and source contacts and hence provides a reduction in cell pitch. Based on two-dimensional simulations, it is demonstrated that the proposed structure provides a BFOM of 22.40 MW/cm^2 with a reduced cell pitch of 3 μm while maintaining a good switching performance.

DEVICE STRUCTURE

The schematic cross-sectional views of the conventional and proposed LDMOS structures implemented on thin SOI are shown in Fig. 1. Both the devices have N$^+$-polysilicon gates with identical gate length of 0.5 μm. As shown in Fig. 1(a), the CLDMOS is a planer device with gate, source and drain contacts on the top of the structure. When a positive gate voltage higher than threshold voltage is applied, a channel is formed in the p-body region under the gate and current flows from drain to source for positive drain voltage. Under off-state, the gate is shorted to the source and applied positive drain voltage causes spreading of depletion region in the N-drift layer. The breakdown of the device occurs when the electric field inside the drift region reaches the critical field of the material. The gate electrode is extended over the drift region to act as field-plate which reduces the electric field on the silicon surface and improves the breakdown voltage. However, the electric field still peaks on the surface at the end of field-plate (point 'A') which limits further improvement in the breakdown voltage. It may be noted that the length of field-plate is among the critical parameters which affect breakdown voltage. At a drift region doping of 2×10^{16} cm^{-3}, the optimized structural parameters of the CLDMOS are listed in Table 1. The total cell pitch of the conventional device is 5 μm. On the other hand, as shown in Fig. 1(b), the proposed TGLDMOS structure has two trenches filled with oxide in the N-drift region on both sides of P-body region. The gate electrodes are symmetrically placed in both the trenches and gate polysilicon is extended over the drift region to work as vertical field-plate. Two channels are formed simultaneously in the P-body along the side-walls of trenches when a positive gate voltage is applied and current flows from the drain to source through the bulk of drift region for positive drain voltages. The presence of two trenches substantially reduces the electric field in the drift region[10]. The trench-structure causes the reduced-surface field (RESURF) effect which leads to full depletion of the drift region resulting large improvement in the breakdown voltage of the proposed device. At breakdown, the peak electric field occurs at point 'C' in the TGLDMOS. It is noteworthy that the TGLDMOS structure reported in [8] has a deep P^+-sinker, spacing between the trench gate and N^+-drain diffusion, and a source field-plate over the trench gate and the drift region. These structural modifications limit the lateral dimensions and hence restrict the further reduction in the cell of the device. On the other hand, the proposed TGLDMOS unit cell contains single drain and source contacts and two gates placed vertically in separate trenches without any horizontal field-plate leading to reduced cell pitch of the device. Moreover, the trench structure folds the drift region in vertical and horizontal directions leading to reduced cell pitch of 3 μm. The optimized device design allow higher drift layer doping of 3×10^{16} cm^{-3} in the TGLDMOS. The breakdown voltage and on-resistance of the proposed structure are affected by various structural parameters whose optimized values are given in Table 1. Note that the proposed device provides 40% reduction in the cell pitch as compared to the CLDMOS. This will result a lower area of fabrication as the drift region resistance is reduced. The proposed TGLDMOS device may be fabricated using a process as reported for other trench-gate structures [10] and [11]. The TGLDMOS structure requires few additional fabrication steps as compared to the conventional device.

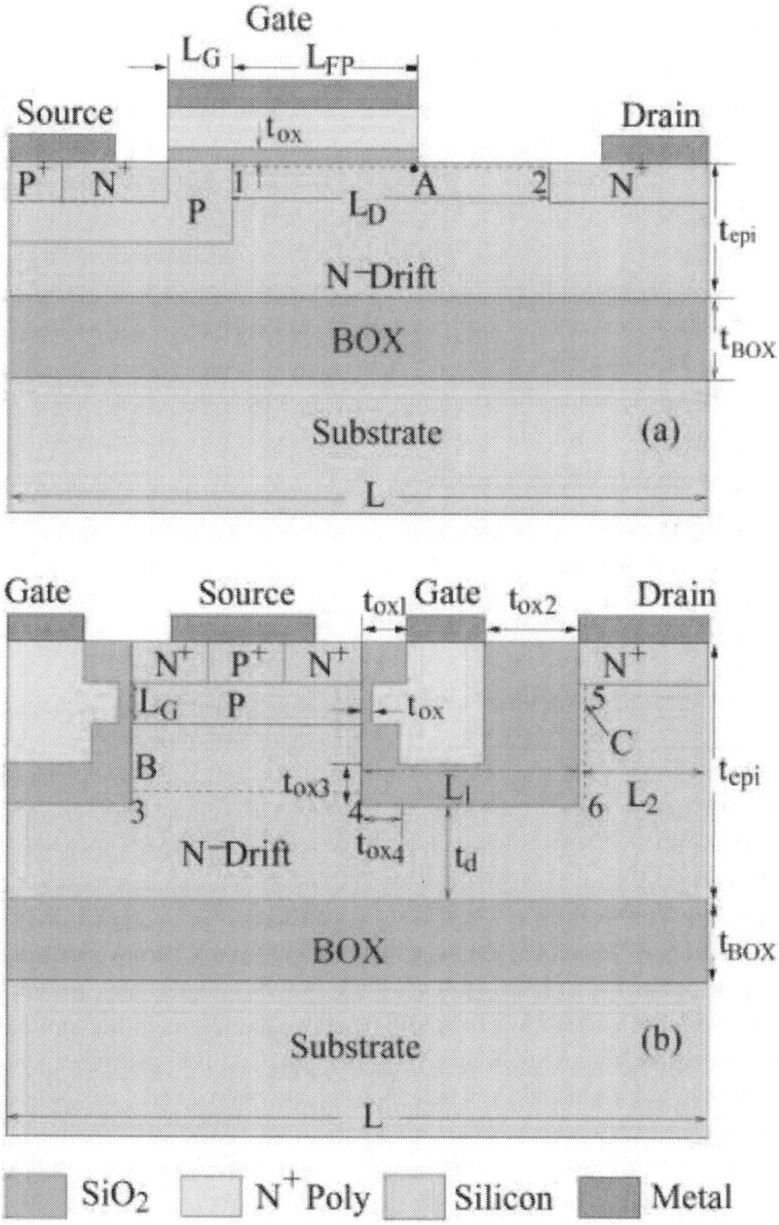

Figure 1. A schematic cross-sectional view of the unit cell of the (a) CLDMOS (b) TGLDMOS.

Table 1. Optimized parameters of the CLDMOS and TGLDMOS.

Parameter	Unit	CLDMOS	TGLDMOS
Cell pitch, L	µm	5	3
Gate length, LG	µm	0.5	0.5
Field-plate length, L_{FP}	µm	1.5	–
Drift region length, LD	µm	2.1	–
Drift region length 1, L_1	µm	–	1.22
Drift region length 2, L_2	µm	–	0.6
Gate oxide thickness, t_{ox}	µm	0.03	0.03
Oxide thickness, t_{ox1}	µm	–	0.18
Oxide thickness, t_{ox2}	µm	–	0.64
Oxide thickness, t_{ox3}	µm	–	0.3
Oxide thickness, t_{ox4}	µm	–	0.25
Epitaxial layer thickness, t_{epi}	µm	0.6	2.25
Drift region thickness, td	µm	–	0.6
Buried oxide thickness, t_{BOX}	µm	0.6	0.7
Drift region doping, Nd	cm^{-3}	2×10^{16}	3×10^{16}
P-body doping	cm^{-3}	8×10^{16}	8×10^{16}

COMPARISON OF SIMULATED RESULTS

To understand the effect of structural parameters on the performance of the CLDMOS and TGLDMOS devices, we have created both the structures in the device simulator (ATLAS) [12]. Two-dimensional numerical simulations were carried out by invoking suitable models in the simulator such as Shockley-Read-Hall model (SRH), Fermi-Dirac model (FERMIDIRAC), concentration dependent mobility model (CONMOB), electric-field-dependent mobility model (FLDMOB), and the Selberherrs impact ionization model (IMPACT SELB). Using the identical simulation models and their parameters for both the structures, the simulated characteristics are compared to evaluate their performance as discussed below:

ON-resistance

Fig. 2 shows the simulated drain characteristics of the CLDMOS and TGLDMOS at gate-source voltage (V_{GS}) of 6 V for low values of drain-source voltages (V_{DS}) such that the devices operate in the linear region. These characteristics are used to determine the on-resistance of the devices. The total on-resistance of an LDMOS is a series combination of channel resistance, drift region resistance, drain and source contact resistances. FromFig. 2, the total on-resistance is calculated as the ratio of drain voltage to drain current and is multiplied by the pitch length to get the specific on-resistance (R_{ON}). The specific on-resistance of the CLDMOS and TGLDMOS are found to be 52 and 42 mΩ-mm^2, respectively resulting 19% reduction in R_{ON} for the proposed device. It may be noted that although, the drain current in the TGLDMOS is lower than that of the CLDMOS but the reduced pitch length leads to smaller R_{ON}. Fig. 3 shows the variation of specific on-resistance with the gate-source bias for the CLDMOS and the TGLDMOS. Initially, asV_{GS} is increased, the on-resistance of both the devices decreases rapidly because of enhancement in the channel conduction. As V_{GS} is further increased, the on-resistance saturates to a particular value indicating that the channel is fully created and the on-resistance is mainly limited by the drift region resistance. We observe that for entire range of V_{GS}, the proposed device exhibits lower R_{ON} as compared to the conventional device. The threshold voltage of the TGLDMOS (1.34 V) is found to be close to that of the CLDMOS (1.58 V). The substrate of both the devices is kept at grounded potential (0 V).

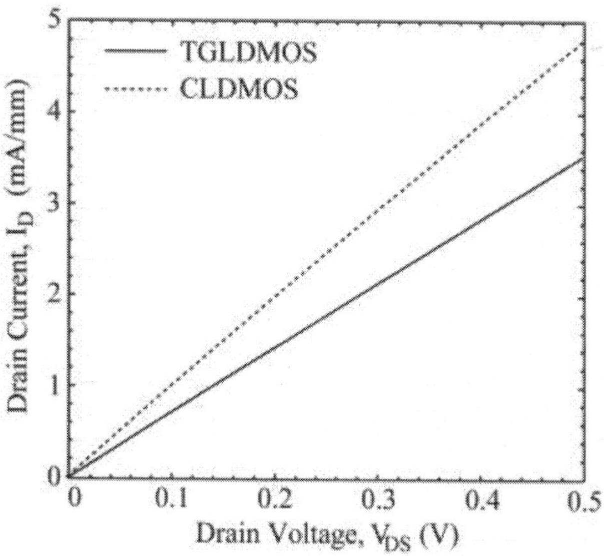

Figure 2. Drain characteristics of the CLDMOS and the TGLDMOS in the linear region for V_{GS} = 6 V.

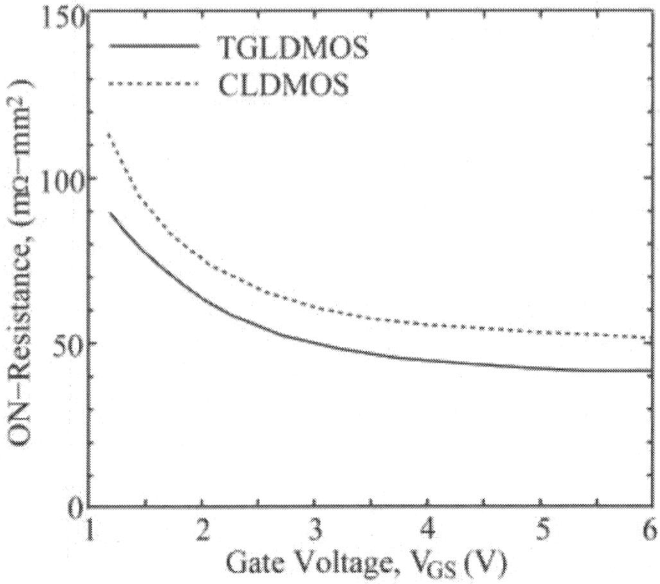

Figure 3. Specific on-resistance variation with applied gate-source bias for the CLDMOS and TGLDMOS.

Breakdown voltage

The simulated breakdown characteristics of the TGLDMOS compared with that of the CLDMOS under off state are shown in Fig. 4. At zero gate bias, there is no path for conduction of electrons from source to drain because the channel is not created in P-base. This results in a normally-off device with the drain voltage mostly supported by the N-drift region. The breakdown voltage is taken as drain to source voltage at which drain current exceeds 0.1 mA/mm. As seen, the breakdown voltage of CLDMOS and TGLDMOS structures are 46 and 97 V, respectively resulting 110% improvement in the breakdown voltage. In order to understand such large enhancement in the breakdown voltage, it is useful to examine the electric field profile inside the drift region of both the devices. Fig. 5 shows the two-dimensional electric field distribution in the drift region of the CLDMOS and TGLDMOS at V_{DS} = 35 V. As seen from the Fig. 5(a), in case of the conventional device, the electric field is crowded on the silicon surface at the end of field-plate (point 'A'). On the other hand, for the proposed structure, the maximum electric field occurs at the bottom edge of the trench (point 'B'). In order to know magnitude of maximum electric field in the drift region of two devices, Fig. 6 gives the electric field variation along the horizontal cut line (1–2 in Fig. 1(a)) located on the top surface of drift region in the CLDMOS and along the cut line (3–4 in Fig. 1(b)) at the bottom edges of two trenches in the TGLDMOS at $V_{DS} = 35$ V. As seen, the peak electric field in the proposed structure is significantly lower than that in the conventional device leading to higher breakdown voltage of the TGLDMOS. It is apparent that in case of the proposed device, the charge increases from source to drain with the applied voltage which modulates the electric field in the drift region resulting enhanced breakdown voltage. Fig. 7 illustrates a two-dimensional view of the electric field in the CLDMOS and the TGLDMOS when the applied drain-source voltage is equal to their breakdown voltages. As seen, the peak of electric field remains at point 'A' in the conventional device whereas it shifts to point 'C' in the proposed device. The peak electric fields at these points can be seen in Fig. 8 which gives the electric variation along the horizontal cut line 1-2 in the CLDMOS and the vertical cut line 5–6 (as in Fig. 1(b)) along the side-wall of trench near the drain. As seen, the peak electric fields in both structures are equal which verify the breakdown of the CLDMOS and TGLDMOS at point 'A' and 'C', respectively. It is noteworthy that the maximum electric field in the TGLDMOS occurs below the semiconductor surface (away from the dielectrics) which typically yields higher critical electric field strength.

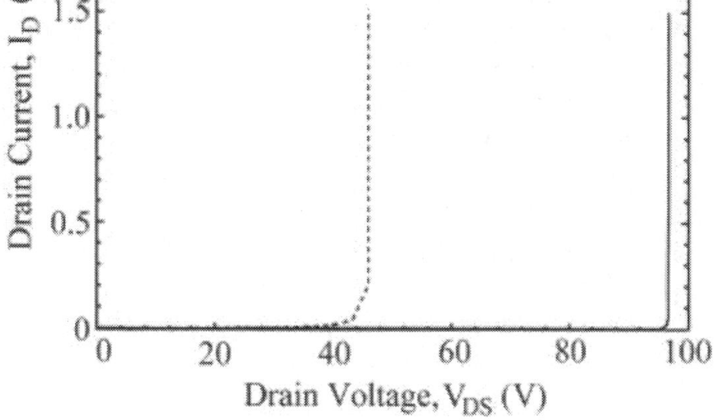

Figure 4. Off-state breakdown characteristics of the CLDMOS and TGLDMOS.

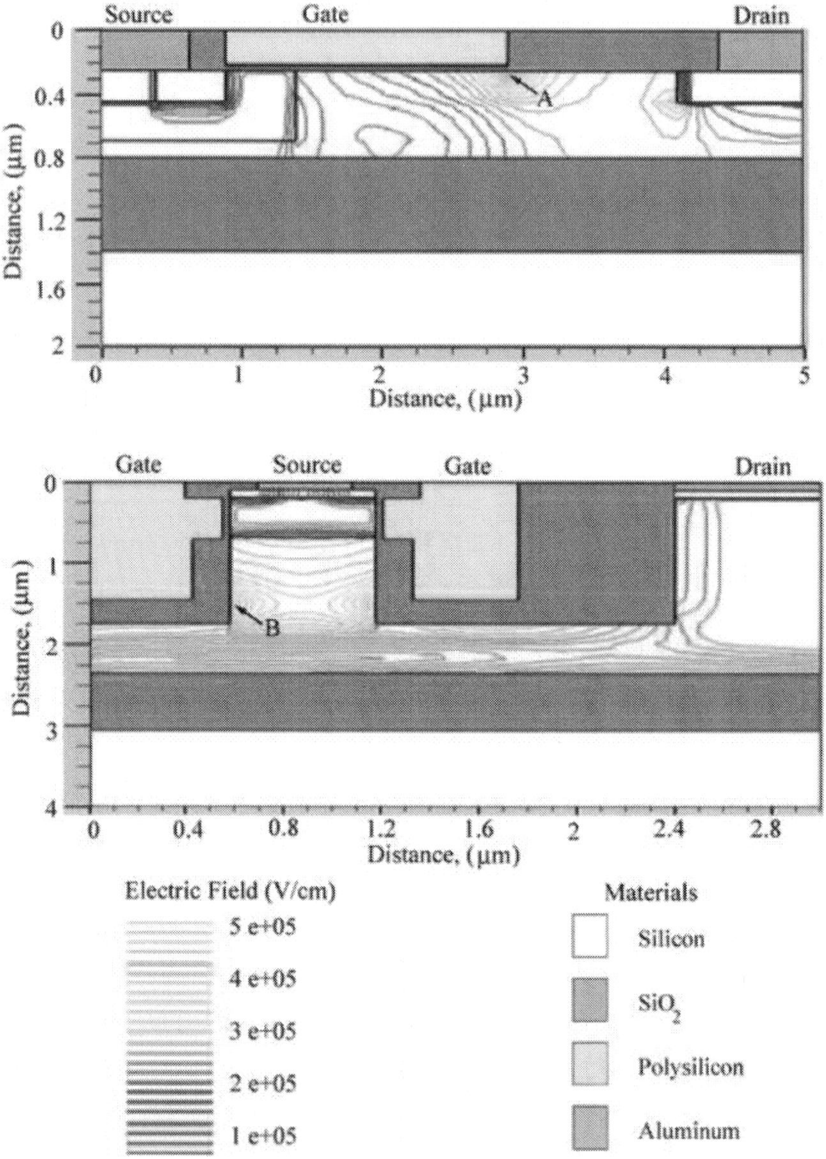

Figure 5. Two-dimensional electric field distribution in the drift region of the (a) CLDMOS and (b) TGLDMOS at V_{DS} = 35 V.

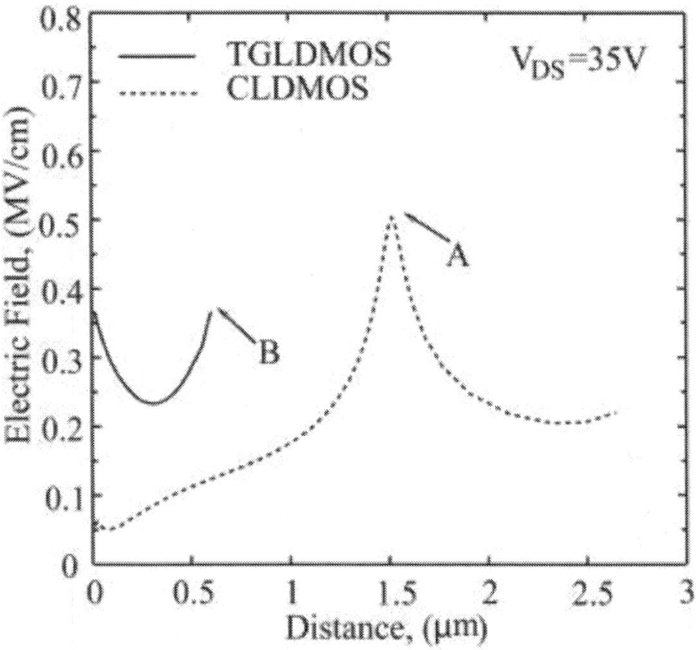

Figure 6. Electric field variation in the drift region of the CLDMOS along cut line 1–2 and the TGLDMOS along cut line 3–4 at V_{DS} = 35 V.

Gate charge analysis

Gate charge (Qg) is the amount of charge required on the gate to turn on the LDMOS for a given on-resistance. This gives an estimation of switching performance of the device. In order to minimize the switching losses of a power LDMOS, the gate charge should be as low as possible. Fig. 9 shows the gate charge curves of the TGLDMOS and the CLDMOS devices. These characteristics are obtained using mixed mode simulations in the device simulator. The circuit used to perform the gate-charge transient is given in the inset of this figure. In these simulations, the device width is kept 4000 μm for both the structures. As seen, for a given V_{GS}, the total Qg of the TGLDMOS is lower than that of the CLDMOS. Moreover, the gate-to-drain charge (Q_{gd}) is taken as the amount of charge that must be input to overcome the "Miller" effect as the drain voltage falls. This occurs during the plateau of the V_{GS} waveform where the voltage is constant. From Fig. 9, theQ_{gd} of the proposed TGLDMOS and CLDMOS devices are found to be 0.62 and 0.88 nC/mm^2, respectively. The proposed device gives 30% lower Q_{gd} as compared to the conventional device. This reduction in Q_{gd} is due to lower gate-to-drain capacitance (C_{gd}) of the proposed device when compared to the CLDMOS as shown in Fig. 10. The CLDMOS structure exhibits higher value of C_{gd} due to presence of a gate field-plate over the drift region with thin oxide (t_{ox}) whereas the TGLDMOS device has thicker oxides (t_{ox2}& t_{ox3}) between the gate and the drift region resulting lower C_{gd}.

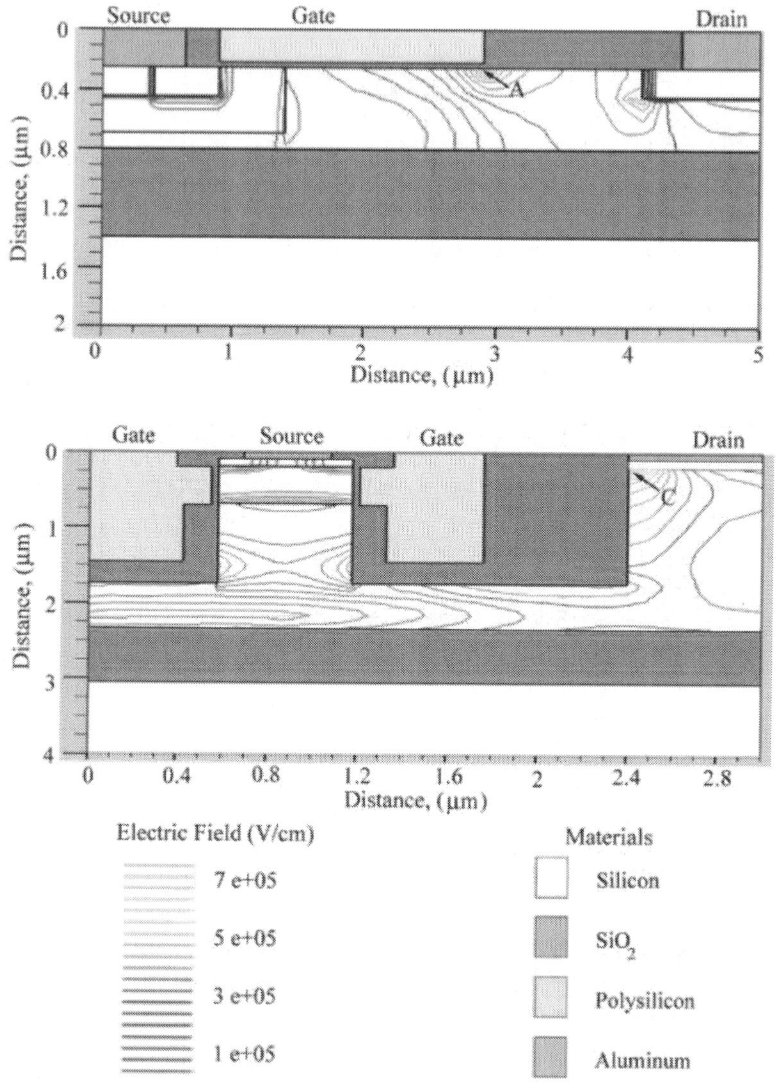

Figure 7. Two-dimensional electric field distribution at breakdown voltage in the drift region of the (a) CLDMOS and (b) TGLDMOS.

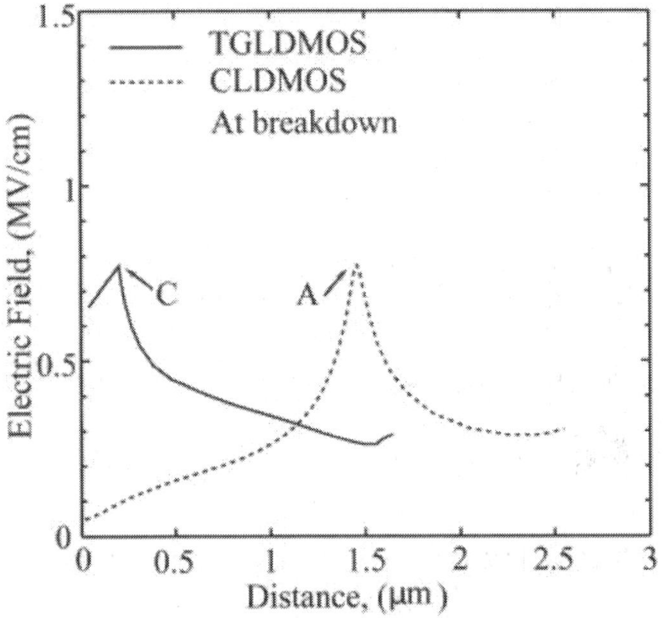

Figure 8. Electric field variation in the drift region of the CLDMOS along cut line 1–2 and the TGLDMOS along cut line 5–6 at breakdown voltage.

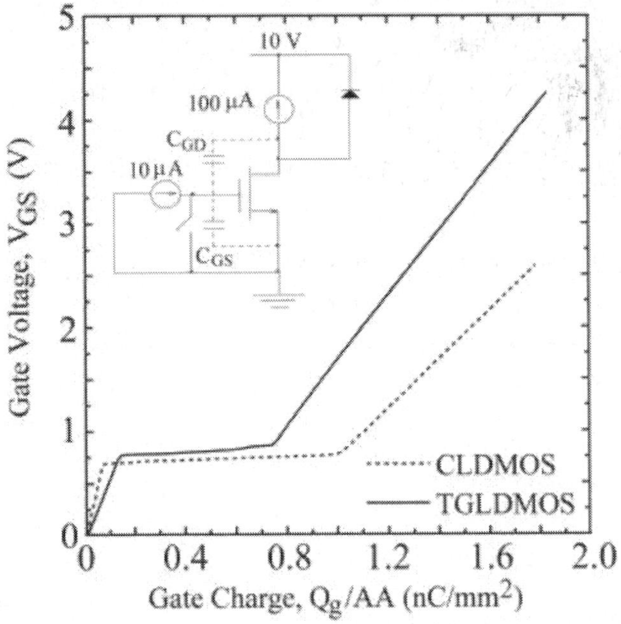

Figure 9. Gate-charge characteristics of the TGLDMOS and CLDMOS.

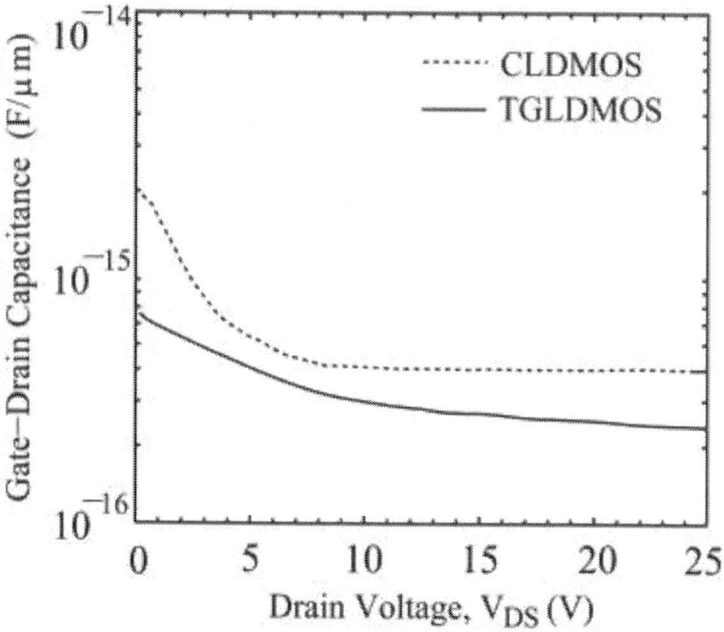

Figure 10. Gate-drain capacitance of the TGLDMOS and CLDMOS.

Figure-of-merit

In the previous subsections, it is observed that the TGLDMOS not only provides lower specific on-resistance but also enhances the breakdown voltage. This indicates that the performance of the TGLDMOS is superior to that of the CLDMOS and yields conditions for a better trade-off between the breakdown voltage and specific on-resistance. From our simulation results, the Baliga's figure-of-merit (BFOM) for the CLDMOS and TGLDMOS is calculated as 4.07 and 22.40 MW/cm^2, respectively. This provides 5.5 times improvement in the BFOM of the proposed device as compared with the conventional device. Further, the dynamic losses figure-of-merit (FOM = $R_{ON}.Q_{gd}$) of the TGLDMOS structure is found to be 26 mΩ-nC as compared to 46 mΩ-nC of the CLDMOS resulting 43% reduction in FOM. A comparison summary of static and dynamic performance of the proposed device with that of the other reported SOI LDMOS devices is given in Table 2. The proposed TGLDMOS structure exhibits significantly higher BFOM than the other counterparts, which demonstrates better suitability of the proposed device for low to medium voltage power applications. The switching performance of the proposed TGLDMOS is comparable to that of the LTDGMOS [9].

Table 2. Performance comparison of power LDMOS devices.

Device structure	V_{BR} (V)	R_{ON} (mΩ-cm²)	BFOM V_{BR}^2/ RON (MW/cm2)	Q_{gd} nC/cm²	FOM $Q_{gd}.R_{ON}$ (mΩ–nC)
CLDMOS	46	0.52	4.07	88	46
TGLDMOS [8]	97.4	1.42	6.68	–	–
LTDGMOS [9]	69	0.37	12.87	63	23
TGLDMOS (This work)	97	0.42	22.4	62	26

DEVICE OPTIMIZATION

In this section, we will investigate the effect of physical parameters on the performance of the TGLDMOS structure. The breakdown voltage and on-resistance are influenced by various device parameters such as the drift region concentration (ND), length of drift region (L_2), drift region thickness (td), the epitaxial layer thickness (t_{epi}), and the oxide thickness (t_{ox2}, t_{ox3}, t_{ox4}).

These parameters are optimized in such a way that the device provides optimal trade-off between the breakdown voltage and the specific on-resistance resulting maximum BFOM. The dependence of the breakdown voltage and BFOM on the doping concentration of drift layer in the TGLDMOS is given in Fig. 11. As the doping concentration increases, the breakdown voltage of the structure enhances due to spreading of the depletion layer over the entire drift region which causes reduction in electric field inside the device. In other words, full depletion of the whole drift region is responsible for higher breakdown voltage. This is due to RESURF effect caused by the presence of trenches in the drift region. However, as the drift layer doping is increased above 3×10^{16} cm^{-3}, the drift region is not fully depleted at the drain side and causes an increased electric field inside the device leading to a rapid drop in breakdown voltage. The structure provides maximum BFOM at an optimum doping of 3×10^{16}. Fig. 12 shows the effect of drift region length (L_2) on the breakdown voltage and figure-of-merit. It is observed that the breakdown voltage of the device decreases as L_2 is varied from 0.40 to 0.85 μm due to increase in the peak electric field at trench bottom. However, the on-resistance of the structure is reduced marginally due to increase in cross-sectional area for flow of the drain current. This leads to an optimum BFOM for a length of 0.6 μm. In the next step, we have optimised the drift region thickness (*td*) by analysing its influence on the breakdown voltage and figure-of-merit as shown in Fig. 13. It can be seen that the breakdown voltage decreases slowly for variation of *td* from 0.20 to 0.65 μm and reduces rapidly thereafter. This is due to change in the electric field distribution in the device with*td*. This thickness also affects the on-resistance of the device due to change in drift region resistance. The optimized value of *td* is obtained as 0.6 μm corresponding to maximum BFOM. Fig. 14 gives a trade-off between the breakdown voltage and on-resistance for change in epitaxial layer thickness (t_{epi}) of the TGLDMOS. Keeping the t_d & t_{ox3} fixed, the variation in t_{epi} scales the length of vertical field-plate (the downward extension of the gate-polysilicon). The breakdown voltage of the structure improves as t_{epi} is increased from 1.6 to 2.6 μm and saturates to a maximum value for further increase in t_{epi}. On the other hand, the on-resistance is also affected by t_{epi} resulting a significant improvement in the BFOM for increase in t_{epi} from 1.6 to 2.25 μm. However, the performance of device degrades after 2.25 μm as optimum value of t_{epi}. The other key dimensions of the structure affecting the breakdown voltage are the oxide thickness t_{ox2}, t_{ox3}, and t_{ox4}. These parameters modulate the electric field distribution inside the device and maximum breakdown voltage is achieved for an optimum value of parameters so that the electric field profile in the device becomes more uniform. Dependence of the breakdown voltage on t_{ox2}, t_{ox3}, and t_{ox4} are illustrated in Fig. 15, Fig. 16 and Fig. 17, respectively. It may be noted that the on-resistance of the device is not affected by these dimensions and hence the trends of BFOM curves are same as that of the breakdown voltage. Simulation results show that the optimum values of oxide thickness are t_{ox2} = 0.65 μm, t_{ox3} = 0.30 μm,t_{ox4} = 0.25 μm. It is to be pointed out that the variation in the oxide thickness, t_{ox1} does not affect the breakdown voltage and the on-resistance of the TGLDMOS. However, the gate-source capacitance increases by reducing the t_{ox1} thickness. The optimised values of all parameters are given in Table 1 which are used in simulations to compare the performance of the proposed device with that of the conventional counterpart.

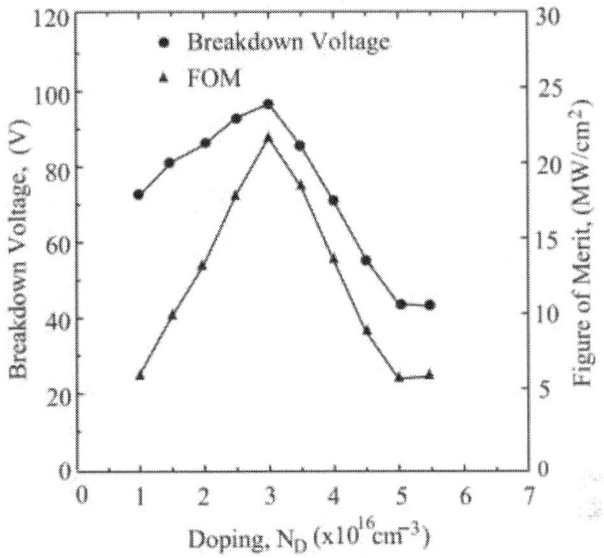

Figure 11. Breakdown voltage and figure-of-merit as a function of drift region doping of the TGLDMOS.

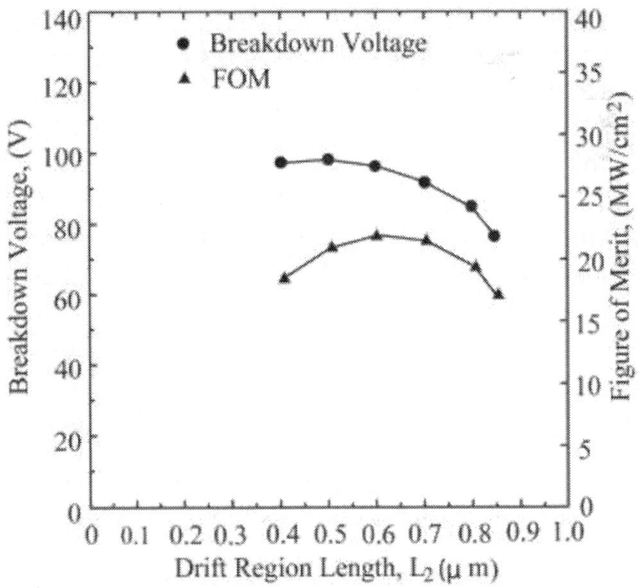

Figure 12. Breakdown voltage and figure-of-merit variation with drift region length (L_2) of the TGLDMOS.

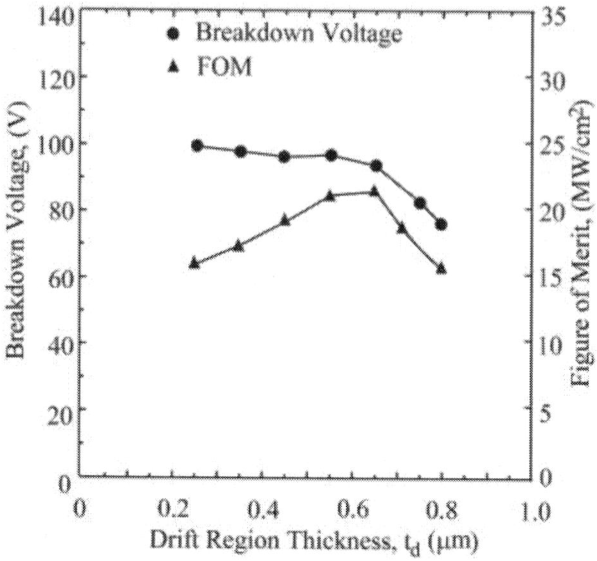

Figure 13. Breakdown voltage and figure-of-merit variation with the drift region thickness (*td*) of the TGLDMOS.

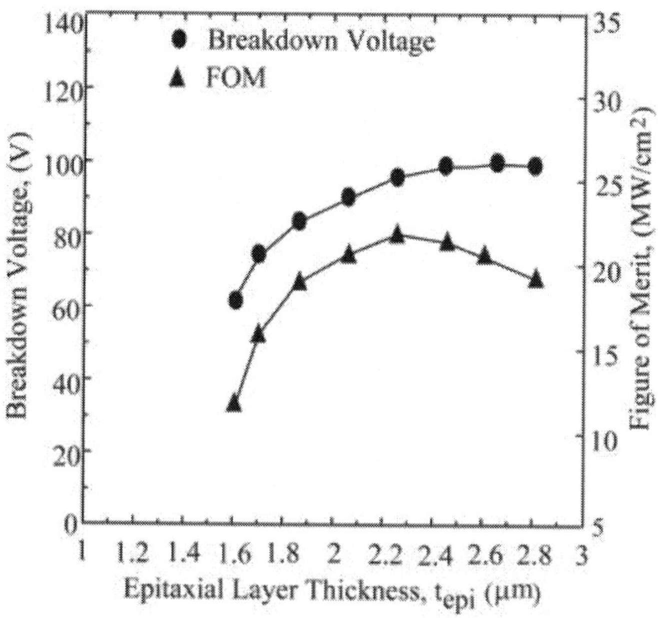

Figure 14. Breakdown voltage and figure-of-merit variation with epitaxial layer thickness (t_{epi}) of the TGLDMOS.

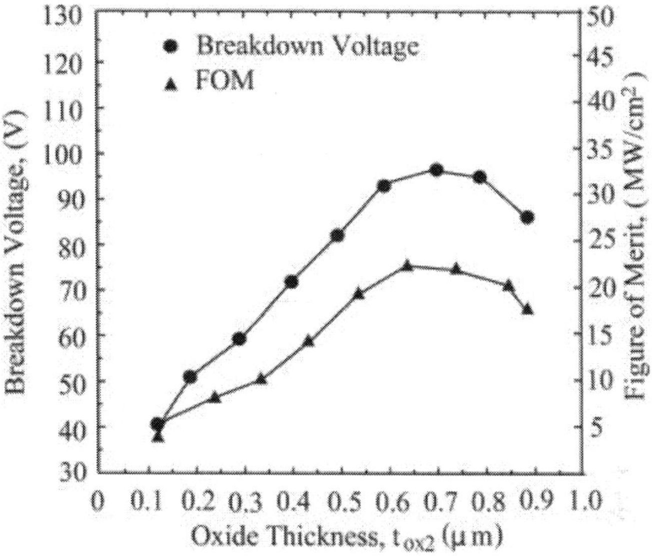

Figure 15. Breakdown voltage and figure-of-merit variation with oxide thickness, t_{ox2} of the TGLDMOS.

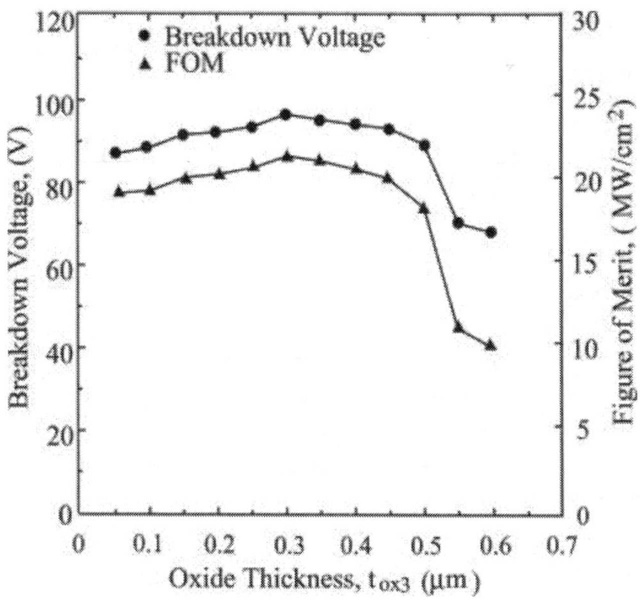

Figure 16. Dependence of breakdown voltage and figure-of-merit on oxide thickness, t_{ox3} of the TGLDMOS.

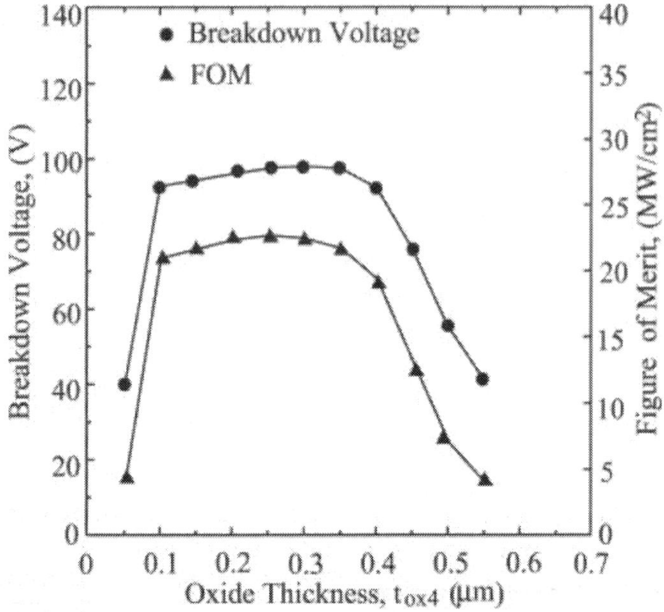

Figure 17. Breakdown voltage and figure-of-merit variation with thickness, t_{ox4} of the TGLDMOS.

CONCLUSION

A power LDMOS structure on SOI is presented which utilizes two vertical gates placed in two trenches on both sides of P-base region. The proposed trench structure suppresses the electric field in the drift region and allow higher drift layer concentration with reduced pitch length leading to a superior performance in terms of maximum breakdown voltage, minimum specific on-resistance, reduced gate-charge, and higher figure-of-merits. Numerical simulations are used to analyse and evaluate the performance of the proposed device and results are compared with that of the conventional device. The device parameters such as drift region concentration (*ND*), drift region length (L_2), drift region thickness (*td*), epitaxial layer thickness (t_{epi}), and oxide thickness (t_{ox2}, t_{ox3}, t_{ox4}) are varied in order to see their influences on the device performance. These parameters are optimised to obtain maximum figure-of-merit. Based on simulations results, it is demonstrated that the proposed structure can provides 110% improvement in breakdown voltage, 19% reduction in specific on-resistance, 30% decrease in gate-to-drain charge and 40%

lower cell pitch when compared with the conventional LDMOS. The Baliga's figure-of-merit of the proposed device is 22.40 MW/cm^2 which is significantly higher than that of the other similar structures reported in literature. These improvements in the performance of the proposed device are achieved at the cost of few additional fabrication steps. Our results show that the proposed device is a suitable power LDMOS for power integrated circuits.

REFERENCES

1. M. Zitouni, F. Morancho, H. Tranduc, P. Rossel, J. Buxo, I. Pags, et al., A new lateral power MOSFET for smart power ICs: the LUDMOS concept, Microelectron. J. 30 (1999) 551e561.
2. M. Rashid, Power Electronics Handbook, Elsevier Inc, 2010.
3. F. Ellinger, Radio Frequency Integrated Circuits and Technologies, Springer, Berlin Heidelberg, 2008, pp. 113e194.
4. F. Schwierz, J.J. Liou, RF transistors: recent developments and roadmap toward terahertz applications, Solid State Electron. 51 (2007) 1079e1091.
5. G. Toulona, I. Cortes, F. Moranchoa, E.H. Bruyered, B. Villardd, W. Torend, Design and optimization of high voltage LDMOS transistors on 0.18 mm SOI CMOS technology, Solid State Electron. 61 (2011) 111e115.
6. W.S. Son, Y.H. Sohn, S.Y. Choi, LDMOSFET with a trench for SOI power integrated circuits, Microelectron. J. 35 (2014) 393e400.
7. T. Erlbacher, A.J. Bauer, L. Frey, Reduced on resistance in LDMOS devices by integrating trench gates into planar technology, IEEE Electron Device Lett. 31 (2010) 464e466.
8. I. Cortes, P.F. Martnez, D. Flores, S. Hidalgo, J. Rebollo, The thin SOI TGLDMOS transistor: a suitable power structure for low voltage applications, Semicond. Sci. Technol. 22 (2007) 1183e1188.
9. Y. Singh, M. Punetha, A lateral trench dual gate power MOSFET on thin SOI for improved performance, ECS J. Solid State Sci. Technol. 2 (2013) 113e117.
10. X. Luo, T.F. Lei, Y.G. Wang, G.L. Yao, Y.H. Jiang, K. Zhou, et al., Low onresistance SOI dual-trench-gate MOSFET, IEEE Trans. Electron Devices 59 (2012) 504e509.
11. N. Fujishima, A. Sugi, S. Kajiwara, Kunio Matsubara, Y. Nagayasu, C.A.T. Salama., A high-density low on-resistance trench lateral power MOSFET with a trench bottom source contact, IEEE Trans. Electron Devices 49 (2002) 1462e1468.
12. Atlas users manual, Device Simulation Software, Silvaco Int., Santa Clara, 2010.

CITATION

Yashvir Singh, Rahul Singh Rawat, High figure-of-merit SOI power LDMOS for power integrated circuits, Engineering Science and Technology, an International Journal, Volume 18, Issue 2, June 2015, Pages 141-149, ISSN 2215-0986, http://dx.doi.org/10.1016/j.jestch.2014.10.004.

Chapter 3

A Method of Multi-Objective Reliability Tolerance Design for Electronic Circuits

Guofu Zhai, Yuege Zhou, Xuerong Ye, Bo Hu

School of Electrical Engineering and Automation, Harbin Institute of Technology, Harbin 150001, China

ABSTRACT

Tolerance design plays an important role in reliability design for electronic circuits. The traditional method only focuses on the consistency of output response. It is not able to meet the needs of increasing development of electronic products. This paper researches the state of related fields and proposes a method of multi-objective reliability tolerance design. The characteristics of output response and operating stresses on critical components are both defined as design objectives. Critical components and their operating stresses are determined by failure mode and effect analysis (FMEA) and fault tree analysis (FTA). Sensitivity analysis is carried out to determine sensitive parameters that affect the design objectives significantly. Monte Carlo and worst-case analysis are utilized to explore the tolerance levels of sensitive parameters. Design of experiment and regression analysis are applied in this method. The optimal tolerance levels are selected in accord with a quality-cost model to improve consistency of output response and reduce failure rates of critical components synchronously. The application in light-emitting diode (LED) drivers indicates details and potential. It shows that the proposed method provides a more effective way to improve performance and reliability of electronic circuits.

INTRODUCTION

Electronic circuits are necessary parts of electronic systems which are widely used in aeronautic and space, industry control, and military fields. Their stability and reliability have a direct and significant effect on system reliability. Practically, performance parameters of components fluctuate around their mean values due to differences of manufacturing techniques and environmental conditions (called as disturbance factors). The fluctuation causes variation and non-conformance of system performance measures. When component parameters deviate from their tolerance levels, it results in soft failure, indicating that the system is functional, but some performance measures do not conform to design specifications.[1, 2 and 3] Especially in fields requiring high reliability, taking a helicopter's automatic driving system for example, the fluctuation of output may generate unpredicted results. Hence, research of tolerance design plays an increasingly important role in electronic circuit design for realizations of high reliability and stability.

Reliability tolerance design for electronic circuits is a topic of great interest. Italian researchers have discussed a tolerance design approach for feedback compensation networks of DC–DC (DC means direct current) switching regulators, identifying performance and stability constraints of crossover frequency and phase margin by means of the Monte Carlo analysis and interval arithmetic computations.[4 and 5] Spanish members have studied output power variation of resonant inverters applied to high pressure sodium (HPS) lamps due to component tolerance. They analyzed lamp aging and tolerance of circuit components, and selected the most suitable value of the resonant network. The result is validated by Monte Carlo analysis using PSpice simulation.[6]However, there are two problems in those approaches. Firstly, the transfer function must be known. The availability is greatly constricted by complicated large-scale circuits, which are difficult to get the system transfer function. Secondly, only stability of output response can be improved. Operating stresses on the components are not paid attention. In fact, working conditions and operating stresses are the most important factors that affect reliability of electronic systems.

With development of circuit simulation technology, Electronic Design Automation (EDA) software has powerful capabilities of modeling and computing. It can assist in performing tolerance design. Presently, tolerance analysis based on EDA simulation has been applied in many electronic systems, such as DC hybrid contactors, driving controllers, and low-pass filters.7 [and] 8 However, they also just analyze factors that cause variation of output characteristics.

The traditional tolerance design method only focuses on consistency of output response. It is a single-objective method. A multi-objective tolerance design method is proposed to improve reliability of electronic circuits. In this research, output response, along with operating stresses on components such as voltage, ripple current, and power loss, are both considered as analysis objects at the same time. Meanwhile, the most important is that operating stresses are the most critical factors that affect reliability of components. Reduction of operating stresses can decrease failure rates of components. So, multi-objective tolerance analysis is not only improving robustness of output response, but also reducing operating stresses on components by tolerance design of parameters. A light-emitting diode (LED) driver is taken as a case to study, determining the optimal tolerance levels of sensitive parameters to increase the qualification rate of a batch of LED drivers, decrease the failure rate, and finally reduce the whole life-cycle cost.

METHOD OF MULTI-OBJECTIVE TOLERANCE DESIGN

The multi-objective tolerance design method focuses not only on output characteristics, but also on operating characteristics of critical components. EDA simulation, sensitivity analysis, experimental design, and multiple regression analysis are adopted into the multi-objective method, which can improve design efficiency greatly. The flowchart of this multi-objective tolerance design method is shown in Fig. 1. In the figure, FMEA means failure mode and effect analysis, FTA means fault tree analysis.

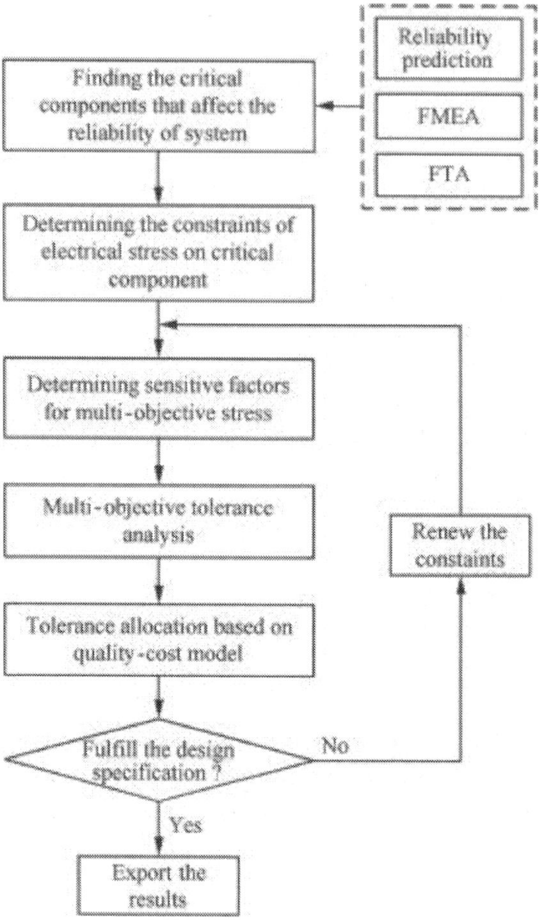

Figure 1. Flowchart of the multi-objective tolerance design method.

Identification of critical components

Critical components are the main factors that affect reliability of a system. There are three methods to determine critical components, i.e., reliability prediction, FMEA, and FTA. Reliability prediction adopts the stress analysis approach to predict failure rate of each component. It is performed under a certain stress. The components with higher failure probabilities are defined as the critical ones. FMEA analyzes all possible failure modes and their effects on the system, determining critical components in accordance with severity and probability. The components with higher risk levels are critical. FTA can identify various failure causes of components and their probabilities. The higher the critical degree is, the greater the probability of system failure is. In this way, those components with greater critical degrees are confirmed as the critical ones. The three methods are combined to identify critical components.

Sensitivity analysis

The purpose of sensitivity analysis is to determine sensitivity parameters causing fluctuation of output response and operating stresses on critical components. Circuit sensitivity is the sensitivity degree of a target response to each design variables, including absolute sensitivity and relative sensitivity. The relative sensitivity is adopted generally to evaluate the degree to which the design factors affect the target response.$f = f(x_1, x_2, \ldots, x_n)$ is defined as the target response in a system, in which $x_1, x_2, \ldots, x_n$ are design variables. Supposing that x_{i0} represents the mean value of a component's parameter, f_0 represents the mean of the target response; then the relative sensitivity $S^f_{x_i}$ can be expressed as

$$S^f_{x_i} = \frac{\mathrm{d}f/f_0}{\mathrm{d}x_i/x_{i0}} = \frac{\partial f}{\partial x_i}\frac{x_{i0}}{f_0} \quad (i = 1, 2, \ldots, n) \tag{1}$$

Orthogonal experiment is an approach that is used in multi-factor tests. It is to select several typical points which are uniform and ordered. Range analysis is applied to analyze the sensitivity of the factors in an orthogonal array.

Tolerance analysis

In a circuit design, the deviation between actual value and nominal value of a component cannot be ignored. Sometimes its effect may be significantly serious. It is essential to study the effect on circuit response because of parameter variation, that is, tolerance analysis. The approach most commonly utilized is the Monte Carlo analysis and the worst-case analysis. The Monte Carlo analysis is a statistical analytical method, which analyzes the response deviation by parameter sampling of components, when the parameter meets certain probability distribution. The result of this method shares the most similarity with actual condition, but consumes much more computing time. The worst-case analysis evaluates circuit performance under a set of worst cases. The result can verify whether the response is acceptable under the worst case. This method has the advantage of reducing analysis time, but the result is too conservative. It just provides the condition under the worst case that is not in accord with most of practical situations. For difference purposes, these two methods are both adopted in the research.

Tolerance allocation

Tolerance allocation means that the system response tolerance is distributed to the tolerances of components. It can be expressed as

$$R_s(R_1, R_2, \ldots, R_i, \ldots, R_n) < R_s^*$$

$$g_s(R_1, R_2, \ldots, R_i, \ldots, R_n) < g_s^*$$

where R_s^* is the tolerance of response, g_s^* the constraint, including cost, temperature, volume, power consumption and so on, R_i the tolerance of design variable *i*. Usually, tolerance allocation is an optimization problem. Therefore, an optimization method is adopted to solve the problem. In this research, the quality-cost model, 9 [and] 10 which is an optimization function for multi-objective tolerance design, is established for the optimal tolerance allocation. The tolerance levels of sensitive parameters are the design variables, and the minimum total life-cycle cost is the final purpose under the condition that the output current and electrical stresses on components fulfill design specifications. The model is consisted of quality loss and manufacturing cost to balance quality and cost increasing due to quality improving. The optimal tolerance levels are the solution of the model, which is solved by genetic algorithm [11] at the lowest total life-cycle cost.

DESIGN AND SIMULATION OF A LED DRIVER

Scheme of the LED driver

The single flyback topology is selected to design a driver for LED lights. Meanwhile, in order to reduce the interference to the grid, a power factor correction circuit is utilized to improve the input current. The schematic diagram is shown in Fig. 2. In the figure, the overvoltage on the field effect transistor VT is reduced by a cushion circuit consisting of a capacitor C_c, a resistor R_c, and a diode D_c.

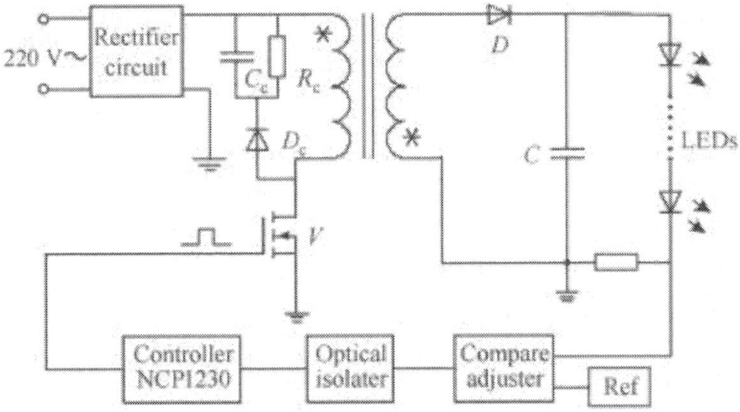

Figure 2. Scheme of the LED driver.

Design specifications of the LED driver

The relationship between bias voltage and current of LEDs is a typical PN junction characteristic. The corresponding forward-bias current increases by orders of magnitude with only a small change in the forward-bias voltage, when the forward-bias voltage exceeds its threshold. Therefore, constant current power supply is selected to drive LEDs.12 [and] 13 The load of the driver is a high power white-light LED, which optimal driving current is determined as 700 mA according to its lighting efficiency. Its light flux increases with current in a linear relation when the forward-bias current is less than 700 mA, while increasing slightly and tending to saturate when the forward-bias current is greater than 700 mA. As a result, the heat dissipation of the LED will dramatically increase, i.e., the light efficiency decreases severely. According to the LED characteristic, the design specifications of the LED driver are given:

1) Output current: (700 ± 15) mA.
2) Operating temperature: −30 60 °C.
3) Qualification rate: greater than 0.95.

Simulation model of the LED driver

The accuracy of analysis based on EDA simulation largely depends on the simulation model. The key component for simulation is the flyback transformer, which directly affects the accuracy of analysis results. The transformer model in PSpice library is ideal and cannot reflect power loss and leakage inductance characteristics. Eqs. (2),(3) and (4) are dynamic equations of the transformer. They are used to establish an accurate flyback transformer simulation model. For tolerance design based on circuit simulations, sensitivity analysis and tolerance analysis are performed based on the steady values of circuit response, such as output current, electrical stresses on components. The steady values of circuit response are closely related to the transformer model. Supposing that R_m represents the transformer core reluctance, let u_1, u_2 and i_1, $i_2$stand for primary and secondary voltage and current respectively, the turn ratio is n_1/n_2, and excitation current is i_{mp}. R_{ac1}, R_{ac2} and L_{l1}, L_{l2} represent the leakage resistance and inductance of primary and secondary winding respectively. R_c is the equivalent resistor of core loss and L_{mp} is the primary magnetizing inductance. The equivalent model of the transformer is shown in Fig. 3.

$$\Phi R_m = n_1 i_1 + n_2 i_2 \quad (2)$$

$$\begin{cases} e_1 = n_1 \dfrac{d\Phi}{dt} = n_1 L_{mp} \dfrac{di_{mp}}{dt} \\ e_2 = n_2 \dfrac{d\Phi}{dt} = n_2 L_{mp} \dfrac{di_{mp}}{dt} \end{cases} \quad (3)$$

$$\begin{cases} u_1 = R_{ac1} i_1 + L_{l1} \dfrac{di_1}{dt} + e_1 \\ u_2 = R_{ac2} i_2 + L_{l2} \dfrac{di_2}{dt} + e_2 \end{cases} \quad (4)$$

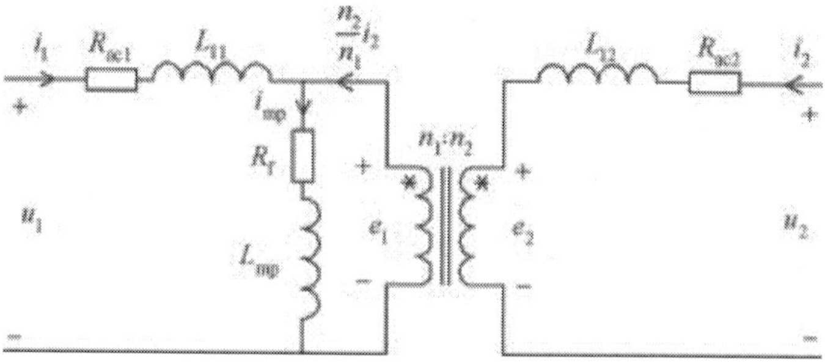

Figure. 3. Equivalent model of the transformer.

The PSpice model of the LED driver is shown in Fig. 4. It can simulate the actual working conditions of the circuit. The implementation of the multi-objective tolerance design with PSpice simulation also has a significant value.

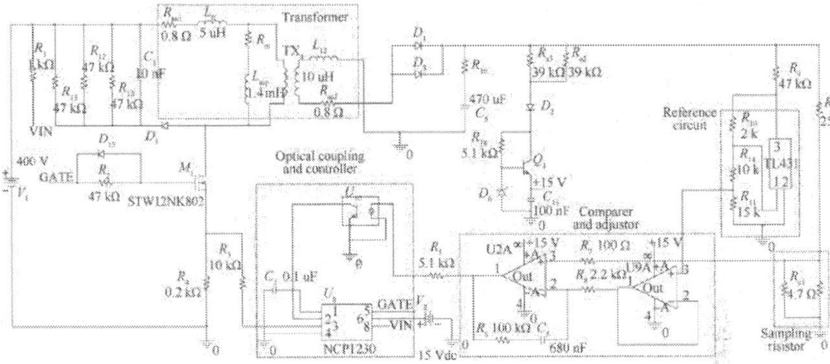

Figure 4. PSpice model of the LED driver.

APPLICATION

The main process is stated in detail in the following sections, using the method of multi-objective tolerance design to analyze the LED driver.

Identification of critical components

The multi-objective tolerance design is different from the traditional one, as it considers the output characteristic of every component that makes up a system. The reliability of the system is improved by means of reducing fluctuations of electrical stress on critical components. According to the reliability analysis, the critical components of the LED driver are metallic oxide semiconductor field effect transistor (MOSFET) Q_3, aluminum electrolytic capacitor C_5, the transformer T_1, and the power rectifier diode D_5. The component failure mechanism shows that decreasing electrical stress can effectively reduce the failure rate and enhance its reliability. This can be achieved by controlling some certain sensitive parameters and related tolerances. The main stresses that cause the critical components to fail are P_M (power consumption on Q_3), P_L (energy dissipation on T_1), V_C (voltage on C_5), I_C (ripple current on C_5), and V_D (reverse-bias voltage on D_5). Therefore, the characteristics of these electrical stresses and output current are all set as the objectives of tolerance design. The constraints of these objectives are determined in accord with the reliability requirements, as shown in Eq. (5), in which $x_1, x_2,\ldots, x_n$ are sensitive parameters.

$$\begin{cases} P_M(x_{M_1}, x_{M_2}, \ldots, x_{M_n}) \leqslant 9 \text{ W} \\ P_L(x_{L_1}, x_{L_2}, \ldots, x_{L_n}) \leqslant 5.5 \text{ W} \\ V_C(x_{C_1}, x_{C_2}, \ldots, x_{C_n}) \leqslant 150 \text{ V} \\ I_C(x_{C_1}, x_{C_2}, \ldots, x_{C_n}) \leqslant 1.65 \text{ A} \\ V_D(x_{D_1}, x_{D_2}, \ldots, x_{D_n}) \leqslant 610 \text{ V} \\ I_{\text{out}}(x_{\text{out1}}, x_{\text{out2}}, \ldots, x_{\text{out}n}) \leqslant 715 \text{ mA} \\ I_{\text{out}}(x_{\text{out1}}, x_{\text{out2}}, \ldots, x_{\text{out}n}) \geqslant 685 \text{ mA} \end{cases} \quad (5)$$

Sensitivity analysis

To find the sensitive parameters, the PSpice simulation is operated to carry out sensitivity analysis for the factors that affect the fluctuations of those electrical stresses. Orthogonal array L_{32} (2^{31}) is selected among the total 27 factors of the LED driver to arrange test order, as shown in Table 1. Then, all factors' relative sensitivity is calculated separately. The results are shown in Fig. 5 and Fig. 6. The sensitive parameters that affect the multi-objective stresses and output current are given in Eq. (6), where the elements in the left column such as U_5, R_2 are sensitive components correspondingly.

$$\begin{bmatrix} 1 & 0 & 0 & 1 & 1 & 1 & 1 & 0 \\ 1 & 1 & 1 & 1 & 1 & 1 & 1 & 0 \\ 1 & 0 & 0 & 1 & 1 & 1 & 1 & 0 \\ 1 & 1 & 0 & 1 & 1 & 1 & 1 & 1 \\ 1 & 0 & 0 & 1 & 1 & 1 & 1 & 0 \\ 1 & 0 & 0 & 1 & 1 & 1 & 1 & 0 \end{bmatrix} \begin{bmatrix} U_5 \\ R_2 \\ R_5 \\ R_{11} \\ R_{14} \\ R_{15} \\ R_{16} \\ C_1 \end{bmatrix} \Longleftrightarrow \begin{bmatrix} P_M \\ P_L \\ V_C \\ I_C \\ V_D \\ I_{\text{out}} \end{bmatrix} \quad (6)$$

Table 1. Orthogonal design factors and levels.

Factor	D_6 (V)	U_5 (V)	R_1 (Ω)	R_{12} (kΩ)	C_1 (nF)	R_2 (Ω)
Level 1	15	2.495	100	47	10	47
Level 2	15.75	2.62	105	49.35	10.5	49.35
Factor	L_1 (μH)	R_4 (Ω)	R_3 (kΩ)	C_2 (μF)	C_{10} (μF)	C_5 (μF)
Level 1	6	0.33	10	0.1	1	470
Level 2	6.3	0.347	10.5	0.105	1.05	493.5
Factor	R_{17} (kΩ)	R_{18} (kΩ)	C_{15} (μF)	R_6 (kΩ)	C_3 (nF)	R_7 (kΩ)
Level 1	39	5.1	10	100	680	1
Level 2	40.95	5.355	10.5	105	714	1.05
Factor	R_{10} (Ω)	R_{14} (kΩ)	R_{11} (kΩ)	R_9 (kΩ)	R_{15} (Ω)	R_{16} (Ω)
Level 1	5.1	10	15	5.1	3.9	4.7
Level 2	5.355	10.5	15.75	5.355	4.095	4.935
Factor	L (mH)	R_5 (kΩ)	R_8 (kΩ)			
Level 1	1.4	10	2.2			
Level 2	1.47	10.5	2.31			

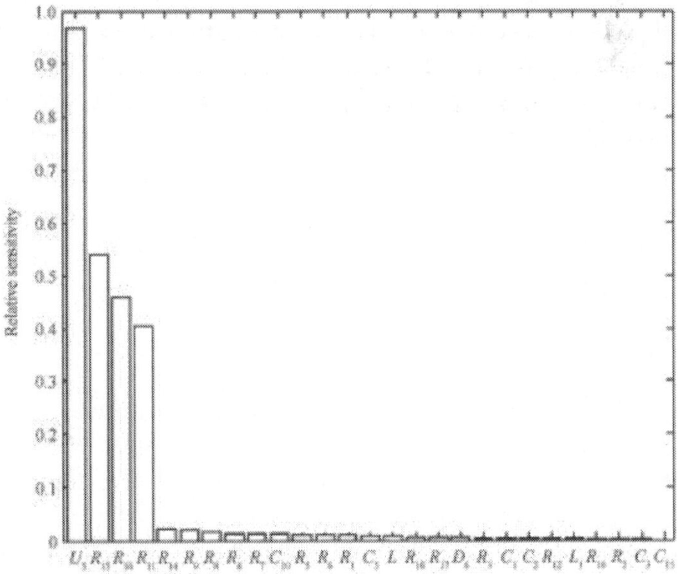

Figure 5. Relative sensitivity of output current.

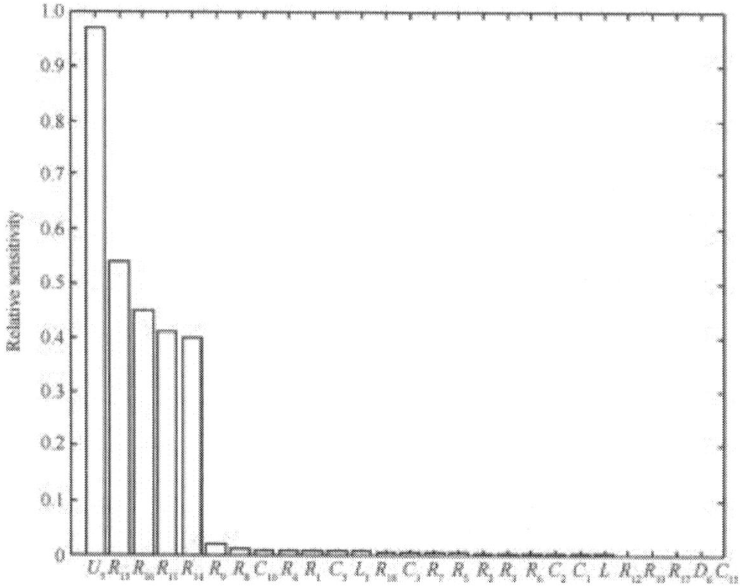

Figure 6. Relative sensitivity of voltage on capacitor.

Tolerance analysis

According to the results of sensitivity analysis, there are eight parameters in total that are sensitive to the multi-objective electrical stresses and output current. They are U_5, R_2,R_5, R_{11}, R_{14}, R_{15}, R_{16}, C_1. In practical applications, the tolerance levels that can be acceptable for the eight parameters are just selected using commercial values. U_5represents the output voltage of the precision regulator TL431, which has three levels: 0.5%, 1%, 2%; resistor R_i has six levels: 0.1%, 0.2%, 0.5%, 1%, 2%, 5%; and capacitorC_i has three levels: 5%, 10%, 20%. Other non-sensitive parameters are kept at the maximum tolerance region, i.e., resistor 5%, capacitor 20%.

Uniform design is adopted to arrange the process of simulation analysis, which can improve the efficiency of tolerance design. Multi-objective electrical stresses are expected to be as small as possible; therefore, only worst-case analysis is performed for these stresses.14 [and] 15 Taking V_C for example, its fluctuations are desired as small as possible. When V_C is still in the range of constraint at the worst case, the result of multi-objective tolerance design fulfills the requirement. U_{12} (6^5) uniform array is selected to perform worst-case analysis. The results are shown in Table 2. Utilizing multiple linear regression method to analyze the results in Table 2, the relationship between V_C and the tolerance levels of sensitive parameters is shown in Eq. (7). Using the Statistical Product and Service Solutions (SPSS) software to perform significance test for the regression equation, it shows that the significance level is close to zero, indicating the significance of the linear relationship between V_C and the tolerance levels.

$$V_C=142.145+149.106t_{U5}+58.854t_{R11}+61.967t_{R14}+82.064t_{R15}+66.676t_{R16} \quad (7)$$

Table 2. Uniform array design.

Test No.	ΔU_5(%)	ΔR_{11}(%)	ΔR_{14}(%)	ΔR_{15}(%)	ΔR_{16}(%)	Target value(V)
1	2	2	0.5	2	5	151.501
2	1	2	5	0.5	0.2	148.372
3	2	0.1	1	5	0.2	149.997
4	0.5	0.2	0.5	0.1	0.1	143.384
5	1	5	0.2	1	0.1	147.573
6	1	0.1	0.2	0.5	5	147.586
7	0.5	1	2	5	1	149.579
8	1	0.2	5	1	2	149.043
9	2	1	0.1	0.1	1	146.61
10	0.5	0.5	0.1	2	0.5	145.192
11	2	0.5	2	0.2	0.5	147.223
12	0.5	5	1	0.2	2	148.001

For the definite purpose of the output current I_{out}, this paper uses the Monte Carlo method to improve the output current fluctuations within the range of (700 ± 15) mA and qualification rate higher than 95%. For example, the result of the Monte Carlo analysis with U_{12} tolerance ± 2%, R_{11} ± 0.1%, R_{14} ± 1%, R_{15} ± 2%, and R_{16} ± 2% is shown in Fig. 7. It can be seen that the output current of the LED driver is close to normal distribution.

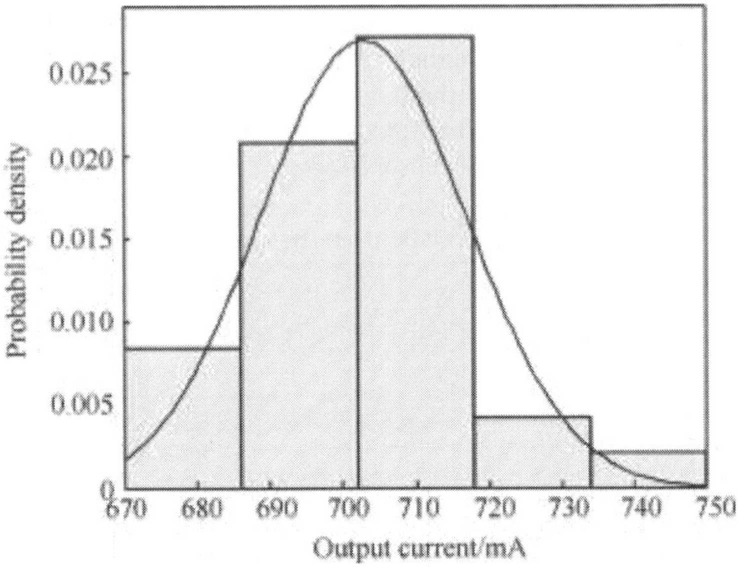

Figure 7. Statistical bar chart of output current.

For distribution I_{out} obeys $N(\mu, \sigma^2)$, with the help of the stress–strength interference model, the qualification rate of the output current is calculated, where x_1 and x_2 are the constraints.

$$P(x_1 < I_{\text{out}} < x_2) = \int_{x_1}^{x_2} \frac{1}{\sqrt{2\pi\sigma}} \exp\left(-\frac{(x-\mu)^2}{2\sigma^2}\right) \mathrm{d}x \tag{8}$$

By means of Eq. (8), the qualification rate of output current in every tolerance combination can be obtained. The regression function between the qualification rateα_{Iout} and the tolerance levels can be built using the multiple linear regression method similarly, as shown in Eq. (9).

$$\begin{bmatrix} V_C \\ P_M \\ P_L \\ I_C \\ V_D \\ \alpha_{I_{out}} \end{bmatrix} = \begin{bmatrix} 149.106 & 0 & 0 & 0 & 58.854 & 61.967 & 82.064 & 66.676 \\ 22.764 & 0 & 0 & 0 & 7.231 & 6.7 & 11.098 & 8.46 \\ 12.193 & 0 & 0 & 1.059 & 5.301 & 5.151 & 7.636 & 6.269 \\ 4.429 & 0.05 & 0.278 & 0 & 1.487 & 0.554 & 1.732 & 1.274 \\ 178.618 & 0 & 0 & 0 & 111.404 & 99.776 & 106.38 & 91.393 \\ -7.888 & 0 & 0 & 0 & -7.657 & -6.593 & -5.843 & -5.843 \end{bmatrix} \begin{bmatrix} t_{U_5} \\ t_{C_3} \\ t_{R_2} \\ t_{R_{11}} \\ t_{R_{14}} \\ t_{R_{15}} \\ t_{R_{16}} \end{bmatrix} + \begin{bmatrix} 142.145 \\ 8.085 \\ 5.157 \\ 1.588 \\ 602.649 \\ 1.1285 \end{bmatrix} \quad (9)$$

Tolerance allocation

Product quality and cost are a pair of irreconcilable contradiction, which means that improving product quality will inevitably lead to increasing cost. Therefore, a quality–cost model is established to balance quality and cost, and to achieve optimization of whole life-cycle cost. The whole life-cycle cost of the LED driver includes two sections: manufacturing cost C_m, and utilizing cost C_u resulted from quality loss.

The manufacturing cost consists of material and component cost and assembly cost, both of which are closely related to tolerance levels of components. The relationship between the manufacturing cost C_m and the tolerance levels of sensitive parameters is set up by curve fitting, as shown in Eq. (10), where C_B, C_{U5}, C_R, and C_C are initial costs of non-sensitive components, the cost of TL431, the cost of sensitive resistor, and the cost of sensitive capacitor, respectively.

$$C_m(\Delta t)=C_B+C_{U5}+\Sigma C_R+C_C \quad (10)$$

The utilizing cost C_u of the LED driver is proportional to its quality loss, which includes both internal and external loss. This LED driver is a non-repairable system, in which the external loss can be ignored while the internal loss is only related to the qualification rate. The qualification rate varies with different tolerance levels of components. What's more, decrease in the qualification rate will cause increase of the utilizing cost. The relationship between the qualification rate and the utilizing cost is shown in Eq. (11).

$$C_u(\Delta t)=K(1-\alpha_{Iout}) \quad (11)$$

where K is the coefficient related to internal loss, and set to 10 for the LED driver. The whole life-cycle cost of the LED driver is the sum of the manufacturing cost C_m and the utilizing cost C_u, as shown in Eq. (12). To find the optimal tolerance level, the genetic algorithm is adopted to solve the function under the minimum cost and the results are listed in Table 3. Meanwhile, the other non-sensitive parameters are kept within their maximum tolerance levels, that is, resistor 5% and capacitor 20%.

$$\begin{aligned} TC(\Delta t) &= C_m(\Delta t) + C_u(\Delta t) \\ &= 201.4464 + 0.000268 t_{U_5}^{-1.538} \\ &\quad + 0.00193 t_{R_{11}}^{-0.6726} + 0.00193 t_{R_{14}}^{-0.6726} \\ &\quad + 0.00581 t_{R_{15}}^{-0.7107} + 0.00193 t_{R_2}^{-0.6726} \\ &\quad + 0.00193 t_{R_5}^{-0.6726} + 0.004 t_{C_1}^{-2} \\ &\quad + 10[1 - (1.123 \\ &\quad - 7.888 t_{U_5} - 7.657 t_{R_{11}} - 6.593 t_{R_{14}} \\ &\quad - 5.843 t_{R_{15}})] \end{aligned} \tag{12}$$

Table 3. Optimal tolerance allocation.

Sensitive parameter	U_5	R_2	R_5	R_{11}
Tolerance(%)	1	5	5	0.1
Sensitive parameter	R_{14}	R_{15}	R_{16}	C_1
Tolerance(%)	0.2	0.5	0.1	20

DISCUSSION AND COMPARISON

The method of multi-objective tolerance design for the LED driver mainly achieves two purposes: one is to raise the qualification rate of the output current and reduce the increase of cost; the other is to make sure that the stresses on critical components are within the specification limit and to improve the quality and stability of the LED driver. These two purposes are verified below respectively.

Comparison of output characteristic

The output characteristic of the LED driver is mainly reflected in the output current. Its design specification is I_{out} = (700 ± 15) mA. The Monte Carlo method is applied in PSpice simulation. Then, the distribution of the output current before and after tolerance design is obtained. Fig. 8 shows the comparison of output current before and after design, via multi-objective tolerance design, the robustness of the output current is significantly improved. The deviation of output current is limited nearly between 680 mA and 720 mA. The actual output current is measured in the experiment, when LED drivers are produced according to the results after design. It is normally distributed I_{out} obeys$N(695.91, 2.726^2)$, as shown in Fig. 9. It demonstrates the validity of this method.

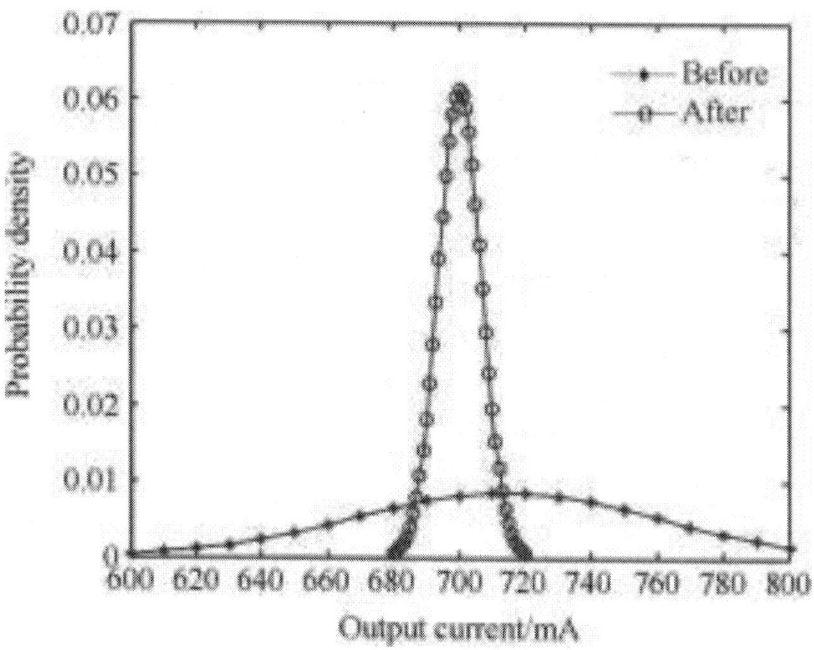

Figure 8. Comparison of output current by simulation.

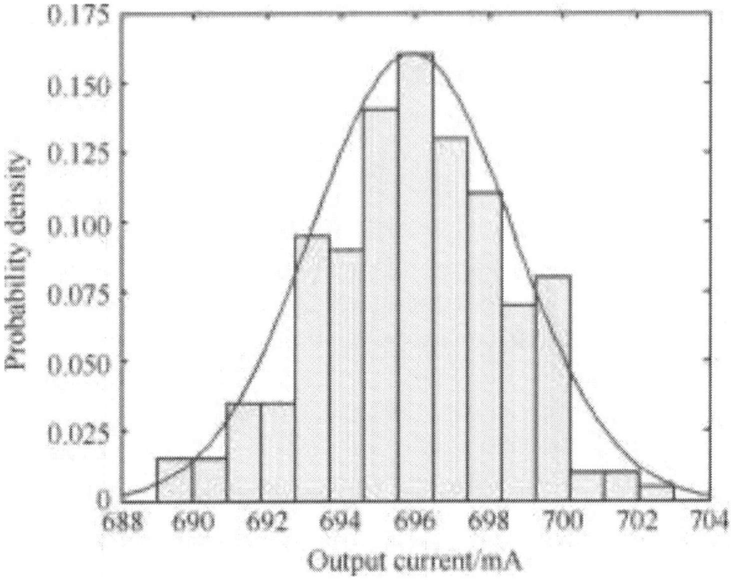

Figure 9. Distribution of output current by experiment.

Fig. 8 and Fig. 9 also show that the simulation result is very close to the actual condition. The output current obtained from simulation fluctuates among 688–713 mA. The mean value and standard deviation are 700.16 mA and 6.511 mA, respectively. The experiment results fluctuate in the range of 689–702 mA. The mean and standard deviation are 695.91 mA and 2.726 mA, respectively. The difference between simulation and experiment results may be due to some components' nominal values deviating from their specifications.

Comparison of electrical stresses

The critical factor that affects system reliability is the reliability of critical components. Fig. 10 shows the comparison of stresses on critical components before and after design. The operating stresses on critical components, such as power loss, voltage, and ripple current, are effectively decreased by the multi-objective tolerance design, thus avoiding the failure due to excessive stress. The mean value and stand deviation are both decreased. The robustness of operating stresses on components can reduce failure rates of components and enhance the life consistency.

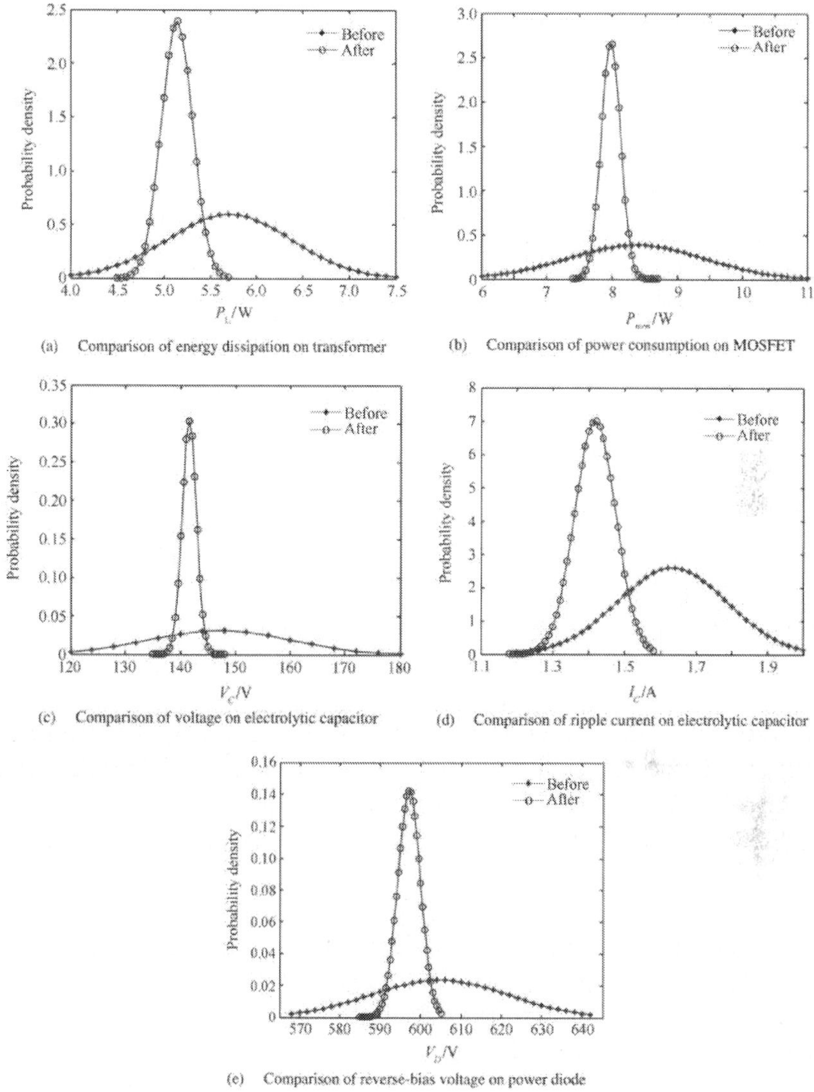

Figure 10. Comparison of stresses on the critical components.

At last, the failure rates of critical components before and after design are predicted according to the approach of stress analysis. The comparison is listed in Table 4. It can be found that the failure rate of some components is reduced to approximately 70%. This result indicates that the reliability of the LED driver is improved greatly and the design specifications are fully satisfied.

Table 4. Comparison of failure rate.

No.	Component	Failure rate (10^{-6}/h) Before design	After design	Ratio
1	MOSFET	3.2638	2.2313	0.6837
2	Electrolytic capacitor	2.8241	1.6063	0.5688
3	Power diode	1.164	0.822	0.7062
4	Transform	0.666	0.636	0.955
5	Transistor	0.3758	0.1508	0.4013
6	Resistor	0.2544	0.2016	0.7925

CONCLUSIONS

This paper proposes a novel multi-objective tolerance design method for electronic systems. It tries to improve the performance of output response and reduce operating stresses on components. The traditional method only focuses on the consistency of output response for a batch of products. It is just an approach to balance cost and performance. The reliability of products themselves is not improved radically. The essential purpose of the proposed method is to improve the working condition and operating reliability of products themselves, and to reduce the cost increase to a maximum extent. It is shown that the operating stresses on some key components and elements are the critical factors that decrease the reliability and life-cycle. From initial design perspective, system reliability can be improved by optimal design. The multi-objective tolerance design is to do such a job.

For a large and complex system, critical components can be identified by reliability prediction, FMEA, and FTA. The key operating stresses on critical components and output response are defined as objectives for multi-objective tolerance design. Sensitive parameters that cause fluctuations of design objectives are determined by sensitive analysis. According to the sensitivity rank, the tolerance levels of sensitive parameters are defined as design variables. Tolerance analysis is performed by the Monte Carlo and worst-case methods. According to the cost and quality loss, a quality-cost model is established. When the whole life-cycle cost is minimal, the tolerance levels are selected.

In the study, the whole process is assisted by PSpice simulation. The proposed method can improve system reliability by tolerance design with high efficiency. It is evaluated by a case study, showing that the operating stresses on a LED driver are reduced significantly and the performance of output current is improved effectively. It solves the problem of improving system reliability from initial parameters and tolerance design.

ACKNOWLEDGEMENT

This study was supported by National Defense Basic Research Program (No.20112060303).

REFERENCES

1. Femia N, Spagnuolo G. Reliable worst-case tolerance design of feedback regulated DC–DC converters by evolutionary algorithms and interval arithmetic. In: IEEE annual power electronics specialists conference. 2002. p. 1728–32.
2. Sharchoudhury A, Kang SM. Worst-case analysis and optimization of VLSI circuit performances. IEEE Trans Comput-Aided Design Integr Circuits Syst 1995;14(4):481–92.
3. Wang SJ, Zhai GF, Chen B, Zhao GL. The research of tolerance analysis and its application of reliability in electronic circuit. Chin J Electron Devices 2002;25(3):296–300 [Chinese].
4. del Casale MDL, Femia N, Lamberti P, Mainardi V. Selection of optimal closed-loop controllers for DC–DC voltage regulators based on nominal and tolerance design. IEEE Trans Industr Electron 2004;51(4):840–9.
5. Petrone G, Spagnuolo G. Tolerance design of controllers for switching regulators. IEEE Trans Aerosp Electron Syst 2004;40(2):661–74.
6. Branas C, Azcondo FJ, Bracho S. Study of output power variation due to component tolerance in LCsCp resonant inverters applied to HPS lamp control. IEEE Trans Industr Electron 2004;51(1):122–31.
7. Wang SH, Sha YT, Zhang H, Zhai GF. Method of reliability tolerance design based on EDA technology and its application on DC hybrid contactor. J Zhejiang Univ Sci A 2007;8(3):426–33.
8. Shi JY, Kang R. Research on method of circuit tolerance analysis based on EDA technology. J Beijing Univ Aeronaut Astronaut 2001;27(1):121–4 [Chinese].
9. Jeang A, Chung CP. Process capability analysis based on minimum production cost and quality loss. Int J Adv Manuf Technol 2009;12(43):710–9.

10. Muthu P, Dhanalakshmi V, Sankaranarayanasamy K. Optimal tolerance design of assembly for minimum quality loss and manufacturing cost using metaheuristic alogorithm. Int J Adv Manuf Technol 2002;12(44):45–9.
11. Busacca PG, Marseguerra M, Zio E. Multiobjective optimization by genetic algorithm: application to safety systems. Reliab Eng Syst Saf 2004;1(72):271–84.
12. Long X, Liao R, Zhou J. Development of street lighting systembased novel high-brightness LED modules. IET Optoelectron 2009;3(1):40–6.
13. Xu X, Wu X. High dimming ratio LED driver with fast transient boost converter. In: 39th IEEE annual power electronics specialists conference. 2008. p. 4192–5.
14. Nassif SR, Strojwas AJ, Director SW. A methodology for worstcase analysis of integrated circuits. IEEE Trans Computer-Aided Design 1986;5(1):104–13.
15. Tian W, Ling XT, Liu RW. Novel methods for circuit worst-case tolerance analysis. IEEE Trans Circuits Syst 1996;43(4):104–13.

CITATION

Guofu Zhai, Yuege Zhou, Xuerong Ye, Bo Hu, A method of multi-objective reliability tolerance design for electronic circuits, Chinese Journal of Aeronautics, Volume 26, Issue 1, February 2013, Pages 161-170, ISSN 1000-9361, http://dx.doi.org/10.1016/j.cja.2012.12.005.

Chapter 4

Failure Analysis of Cracked Brass Turrets Used in Electronic Circuits

Nirmal Ninan, S.V.S. Narayana Murty, Sushant Manwatkar, P. Ramesh Narayanan*

Vikram Sarabhai Space Centre, Indian Space Research Organization, Trivandrum 695 022, India

Highlights

- Failure analysis of cracked brass turrets is discussed.
- Cracking of these critical components can have significant implications to the on-board electronic systems in aerospace applications and therefore should be clearly understood.
- This work describes crack propagation mechanism during cold working of free cutting brass with varying major alloying elements.
- The optimum composition of the alloying elements for obtaining crack-free terminals during cold working is also discussed.

ABSTRACT

Copper alloys are widely used in electrical circuits mainly due to their electrical conductivity. In electrical circuits, solder terminals are used to provide multiple electrical connections and swage type turret terminals are preferred owing to their high reliability under dynamic environments where the electrical systems are subjected to higher vibrations and line currents, as in the case of satellite launch vehicles. Owing to high machinability, free-cutting brass is preferred in the manufacture of such small sized terminals, which will be cold worked/swaged to the printed circuit boards. Improper selection of the composition of free-cutting brass can lead to cracking of the terminals, thereby affecting the reliability of the system. This paper describes cracking of turret terminals made of free cutting brass during swaging operation. The crack propagation tendency during cold working of brass turret terminals made from different compositions is studied through optical and scanning electron microscopy. Finally, composition for obtaining crack free terminals during swaging is suggested for applications involving cold swaging.

INTRODUCTION

Copper and its alloys are widely used for electrical wire and cable, electrical contacts and various other parts that are required to conduct electrical current. Copper is a relatively soft and malleable metal with excellent formability, making it ideal for non-structural applications. Addition of alloying elements to copper strengthens it and forms copper alloys, including brasses, phosphor bronzes and copper nickels. Miniaturization of electronic devices and components has benefited from the high strength and moderate to high conductivity offered by the copper alloys.

Brasses are alloys of copper and zinc and are extensively used in electrical applications due to their physical properties especially their electrical conductivity. They are available in different grades such as alpha brass, beta brass and alpha + beta brass depending on the composition. These alloys exhibit good strength and ductility and are easily cold worked and their properties improve with increased zinc content up to 35% [1]. Brasses containing between 32% and 39% zinc exhibit excellent hot working characteristics even though cold workability is inferior. Brasses containing more than 39% zinc, such as Muntz metal, have high strength and lower ductility at room temperature than alloys with less zinc. Brasses are known for their ease of fabrication by drawing, high cold-worked strength and corrosion resistance.

Miniaturization and weight considerations urge for the requirement of smaller electrical terminals and therefore, free-cutting brass is highly preferred choice because of its excellent machinability. In addition to the basic alloying elements copper and zinc, presence of lead in small quantities gives the material its highest machinability rating. Lead does not form solid solution with copper and zinc at low quantities and it will be distributed in the matrix as globules, which increase the machinability of the alloy. The specification of free cutting brass as per ASTM B16 [2] is as follows: copper 60–63%, lead 2.5–3.0%, impurities 0.35 and zinc–remainder (all in weight percentage). The amount of zinc content affects the equilibrium phases of the alloy and subsequently its cold workability and related properties like hardness.

Based on the specified standard, the amount of zinc can vary from 33.65% to 37.15%. In this band of zinc content, there exists a transition from an alloy with a homogenous α (alpha) phase to an alloy with heterogeneous α + β phase alloy. Even though the presence of β aids in the machinability of the alloy, the hardness and the BCC crystal structure of the β phase will affect the cold workability of the alloy [3]. The volume fraction of the harder phases directly results in the cracking of the alloy during cold working. This paper details the effect of β phase in the free-cutting brass alloy to be cold worked at room temperature and the severity of the cracks formed. The propagation of the cracks during cold working of the material is also discussed.

EXPERIMENTAL

Turret terminals obtained from different sources having different compositions of the alloying elements as per the ASTM B16 standard were tested for their cracking tendency during swaging. The terminals were coated with silver with 10–20 μm thickness for better solderability. The specimens were annealed at 380 °C and furnace cooled. The chemical composition of the materials is confirmed through inductively coupled plasma technique. The dimensions of the specimens are as shown in Fig. 1. The specified dimensions are for the terminal to be swaged on a 1.6 mm thick epoxy material Printed Circuit Boards (PCB).

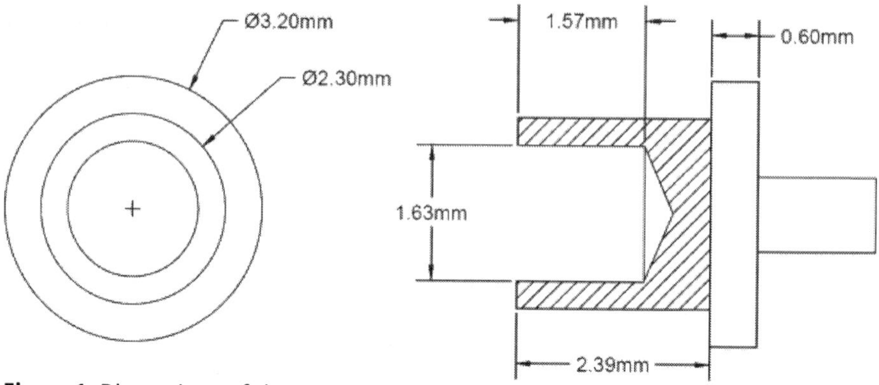

Figure 1. Dimensions of the terminal specimens.

Swaging/cold working method

A manual terminal swager (Cambion 435-3803-01-00-00) is used for the swaging operation [4] of the specimens. The scheme of swaging is shown in the Fig. 2.

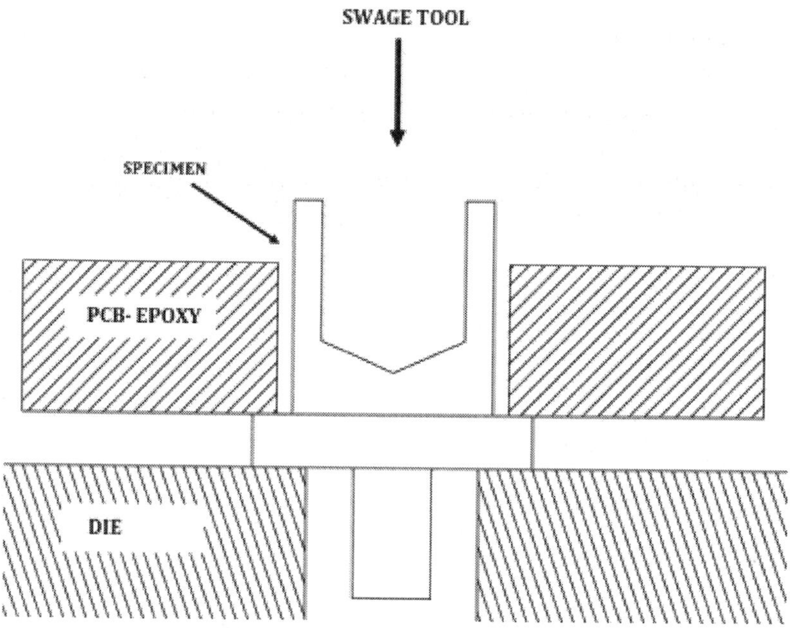

Figure 2. Scheme for cold working the terminal.

Cracked and successfully swaged terminals were photographed using a digital camera and are shown in Fig. 3. Fig. 3(a) shows the photograph of terminals swaged to PCB for testing swageability with the top photograph showing back side and bottom photograph showing the front side of the swaged turret. Fig. 3(b) shows the close up of the terminal with severe edge cracks along with successfully swaged terminals for the purpose of comparison.

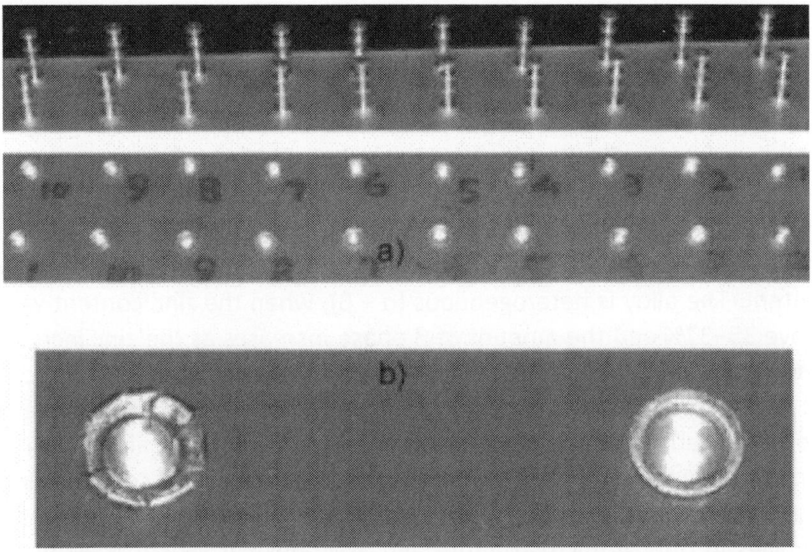

Figure 3. (a) Photograph of terminals swaged to PCB for testing swageability and (b) close up of the terminal with severe edge cracks (left) and successfully swaged terminal for comparison (right).

Metallographic sample preparation

For microscopic examination, the specimens were mounted using a cold setting resin and subjected to conventional metallographic polishing using different grits of silicon carbide papers starting from 220 grit to 600 grit. After a thorough wash, they were subjected to polishing using levitated aluminia and a finish polish of diamond lapping to get a scratch free, mirror finish surface. The freshly polished specimens were then chemically etched using a solution of 2 g of $K_2Cr_2O_7$ in 8 ml H_2SO_4 with 100 ml H_2O to reveal the microstructure. They were examined using an optical microscope (Olympus PMG3) and scanning electron microscope (Carl Zeiss EVO50) and photomicrographs were recorded.

Microhardness

The microhardness of the specimens was measured on freshly polished and etched specimens using a Clemex make automatic microhardness analyser at a load of 100 g. The microhardness values are reported in VPN.

RESULTS AND DISCUSSION

The terminals were analyzed for their microstructure through optical and scanning electron microscope. As the zinc content varies from 34% to 37% in the alloy, significant variations in the microstructure with respect to the phases and the volume fraction of the equilibrium phases exist. The specimen containing less than 35% zinc content is a homogenous α-phase alloy and the precipitation of β phases occurs beyond 35% zinc content. The alloy is heterogeneous (α + β), when the zinc content varies above 35–37% and the amount of β phase increases as the zinc increases monotonically affecting the properties of the free cutting brass especially its cold-workability. Fig. 4(a and b) shows the optical microstructures of the turret terminal containing less than 35% zinc content along with the corresponding SEM photomicrograph. The microstructure contains single phase homogeneous mixture of the solid solution with lead seen as dark particles in optical microstructure and as bright particles in SEM due to their higher atomic number.Fig. 5(a and b) shows the optical photomicrographs and scanning electron micrographs of the material containing 36% zinc content. Presence of beta phase can be clearly seen. In these microstructures also, lead is distributed as globules in the matrix as can be seen by bright particles. Fig. 6(a and b) shows the optical photomicrograph and scanning electron micrograph of the material which contains zinc content of 37 wt.%. It can be clearly seen that the volume fraction of the beta phase is higher in Fig. 6 and it increases with increase in the zinc content.

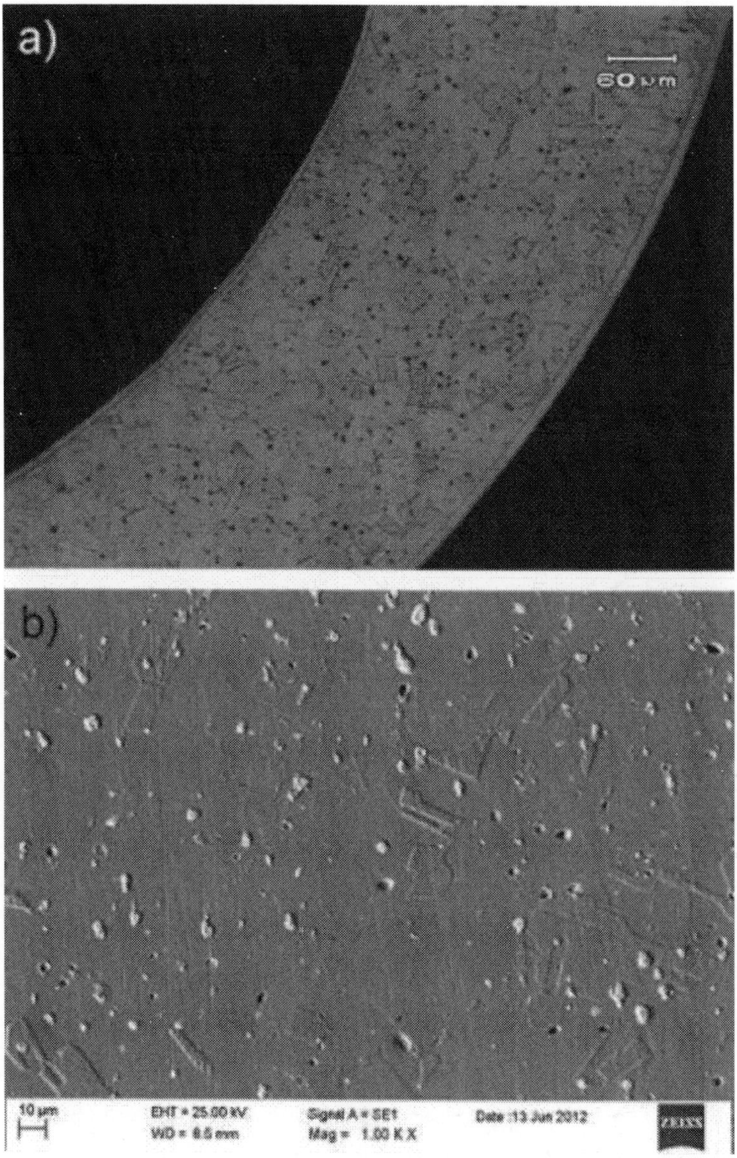

Figure 4. (a and b) Optical microstructure of the turret terminal containing less than 35% zinc content (a) and corresponding scanning electron micrograph (b). The microstructure contains only single phase (homogeneous). Lead can be seen as dark particles in optical microstructure (a) and as bright particles in SEM (b).

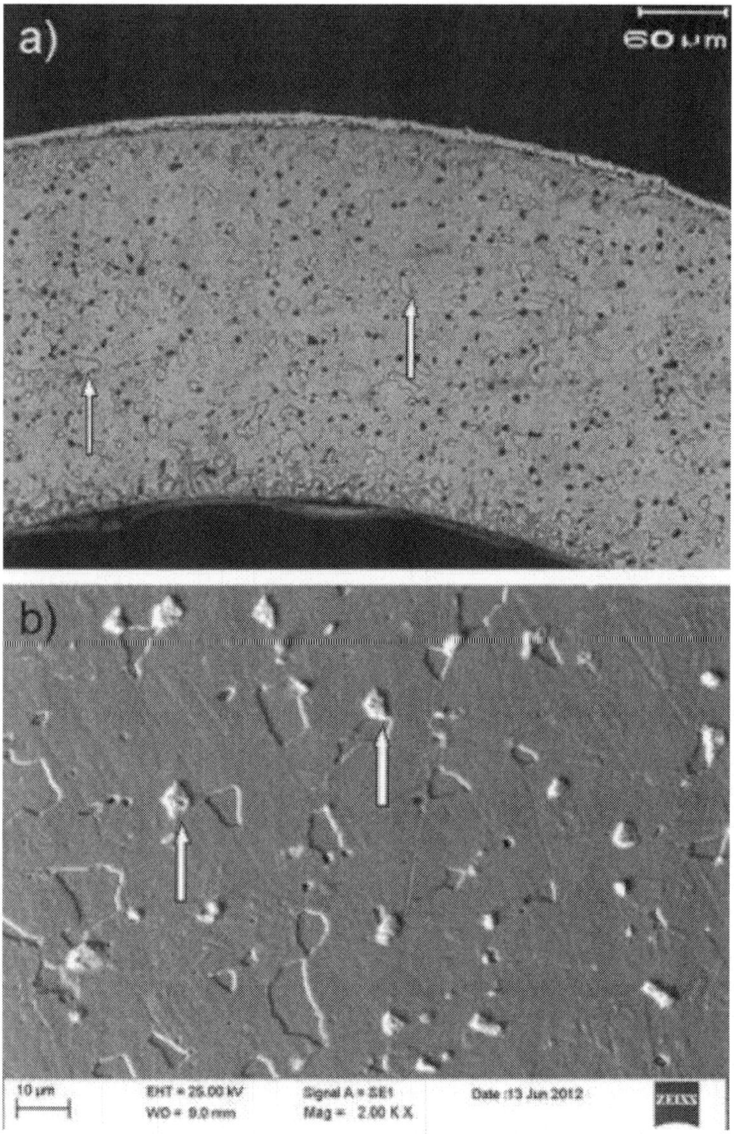

Figure 5. (a and b) Optical photomicrograph (a) and scanning electron micrographs (b) of the heterogeneous microstructure of material containing 36% zinc content. Presence of beta phase can be seen (shown with arrows). Lead is distributed as globules (shown with arrows) in the matrix as can be seen by bright particles in (b).

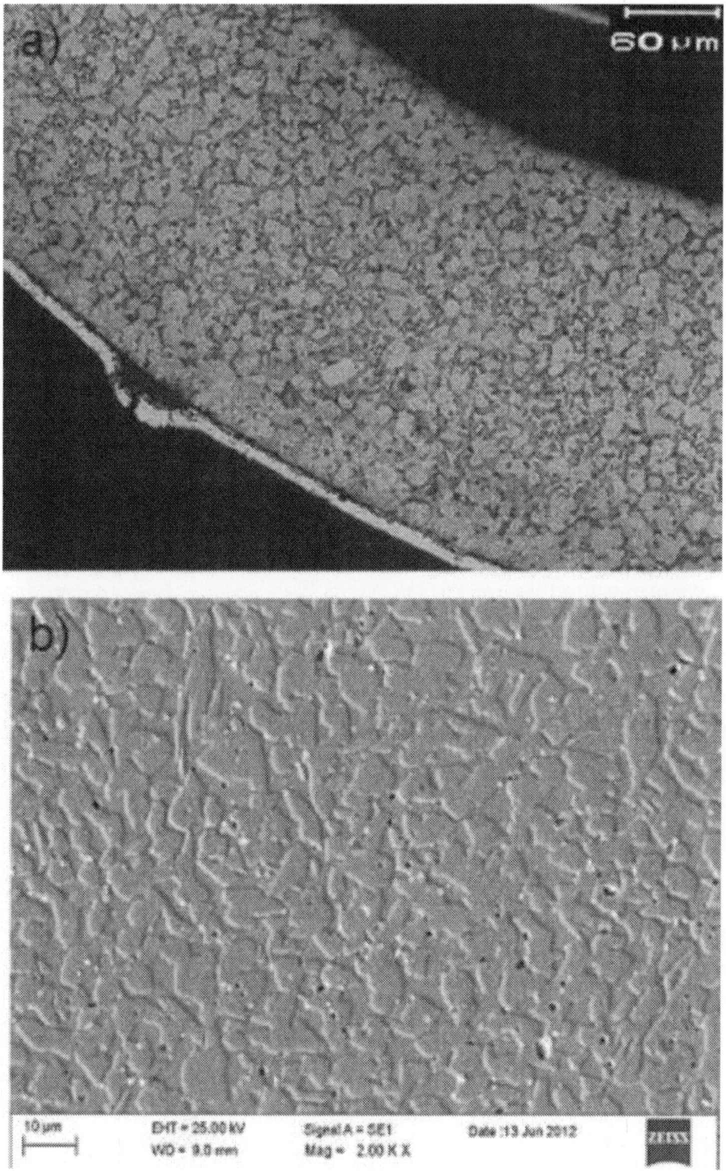

Figure 6. (a and b) Optical photomicrograph (a) and scanning electron micrograph (b) of the heterogeneous microstructure of the material which contains zinc content 37% and above. The volume fraction of the beta phase is higher and it increases with the zinc content.

Cold workability and crack propagation

Copper has high solubility for elements such as nickel, zinc, tin and aluminum. The solid solution alpha (α) phase is responsible for the high ductility exhibited by copper alloys. Alloying additions beyond the solubility limit result in the beta (β) phase, which exhibits a body-centered cubic (bcc) structure. This β phase has high temperature stability, and alloys that exhibit α+β structure have excellent hot forming capability. Hot working, a standard step in alloy manufacture, breaks down the dendritic microstructure present in castings. Some alloys, such as the high zinc brasses, bronzes and nickel silvers, generally exist in two phase α+β condition. Their ability to be cold worked is limited, and they are generally supplied in an extruded or lightly drawn temper condition close to their final form and shape. Fabrication of components is then carried through hot forging and/or machining. Some extruded copper alloy rod and bar products exhibit the dual α+β phase structure which shows some solubility of any deliberate lead additions that, in turn, enhances the hot workability but restricts the cold ductility.

The cold workability of free cutting brass considerably decreases as the amount of zinc content increases or as the amount of beta β phases increases. Cracks were found to develop in the material during cold working which has a heterogeneous microstructure. The intensity and the amount of cracks developed depend on the volume fraction and hardness of the β phases present.

The β phase increases the hardness of the material and thereby increasing the brittleness which directly results in the poor cold workability of the free cutting brass at room temperature. The harder β-phase has the BCC structure while α-phase takes up the FCC structure [2]. Hence the cold workability degrades as the β content increases in the case of α + β alloy [2]. The propagation of cracks in the free cutting brass occurs when the toughness of the material is less than the energy carried by the crack during the time of load application. The toughness is related to the hardness of the material. Higher hardness of the material makes the material more brittle and less tough.

During the cold working of free cutting brass having a heterogeneously phased microstructure, the cracks that are developed from the outer circumference of the specimen, start to propagate when the material is deformed by the gradual load applied. High stress concentrations caused due to the manufacturing defects like tool marks, dents will lead to the initiation of the crack.

Fig. 7(a and b) shows the photomicrographs of cold worked specimen containing more than 37% zinc. Intense edge cracks can be seen on the swaged portions originating from the OD side of the turret and propagating toward the ID side. Fig. 8(a and b) shows the SEM micrographs of the specimens with zinc content varying from 36% to 36.5%. Crack formation is observed albeit to a lesser extent in specimens with lesser zinc content, because of less β content in the microstructure. Fig. 9 shows the SEM photomicrographs of specimen with less than 35% zinc content, showing excellent deformation of the turret head without any edge cracks. Fig. 10 shows the optical photomicrographs of the propagation of the crack through the microstructure. The path of the crack can be seen and it passes through the beta phase. Fig. 11 shows the optical photomicrographs of the propagation of crack through the microstructure. The crack can be seen to be propagating through the harder beta phases present in the microstructure clearly.

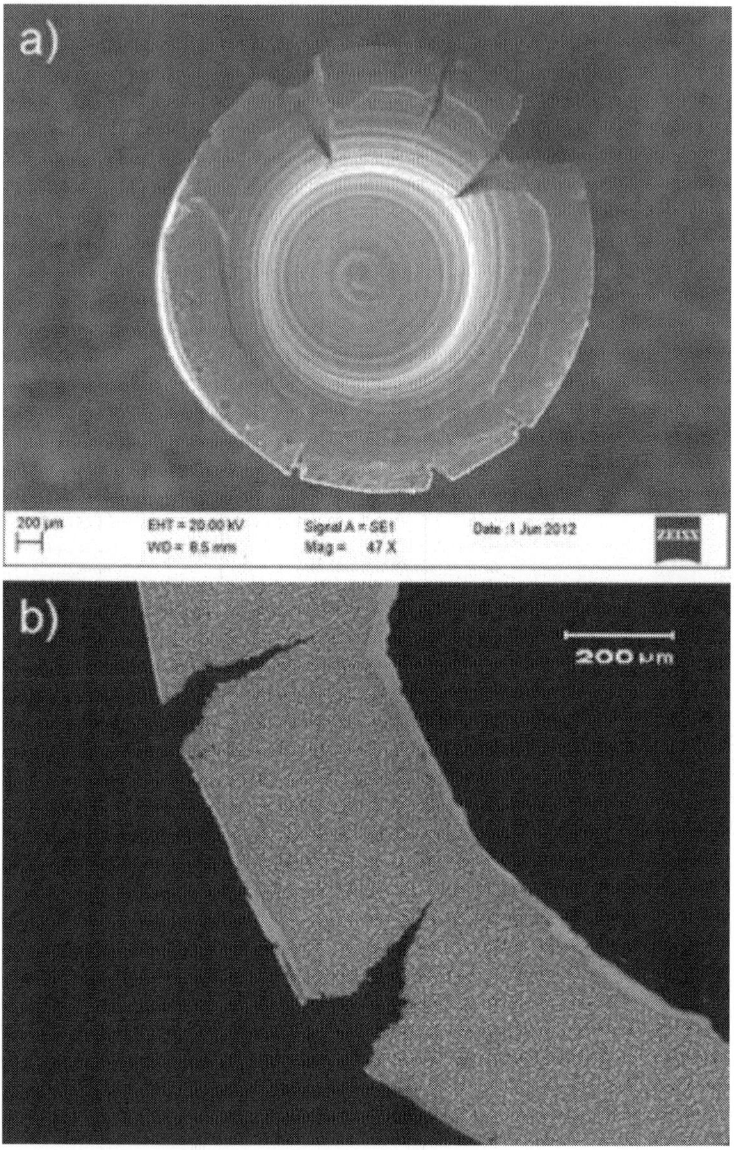

Figure 7. (a and b) Scanning electron micrographs of cold worked specimen containing more than 37% zinc. Intense edge cracks are developed on the swaged portions originating from the OD side and propagating to the ID.

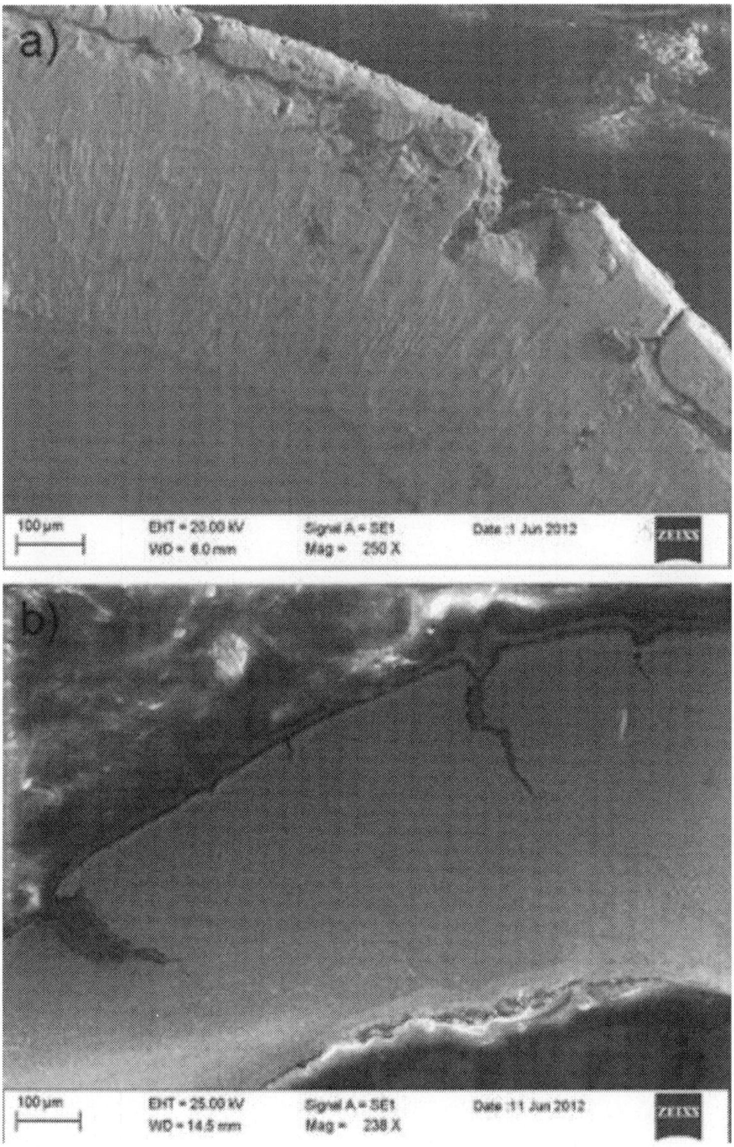

Figure 8. (a and b) Scanning electron micrographs of the specimens with zinc content varying from 36% to 36.5%. Crack formation is observed albeit to a lesser extent in specimens with lesser zinc content, because of less β content in the microstructure.

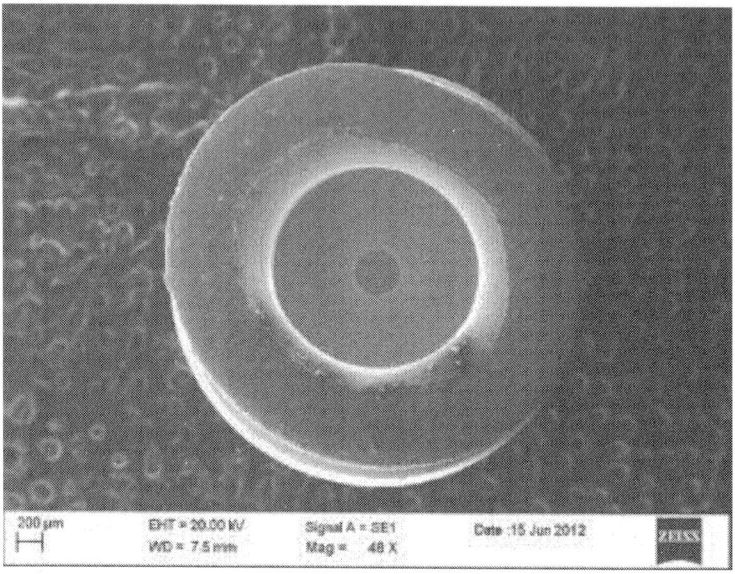

Figure 9. Scanning electron micrograph of specimen with less than 35% zinc content, shows excellent deformation (cold-workability) without the development of any edge cracks.

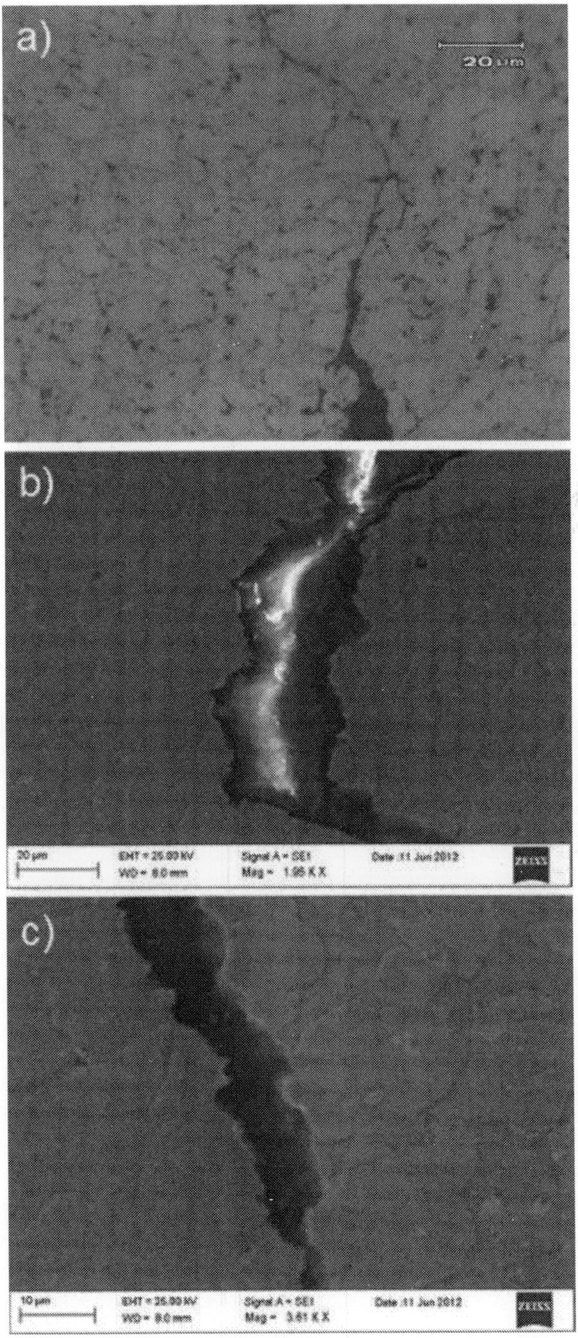

Figure 10. Scanning electron micrographs of the propagation of the crack through the microstructure (OM and SEM). The path of the crack can be seen and it passes through the beta phases.

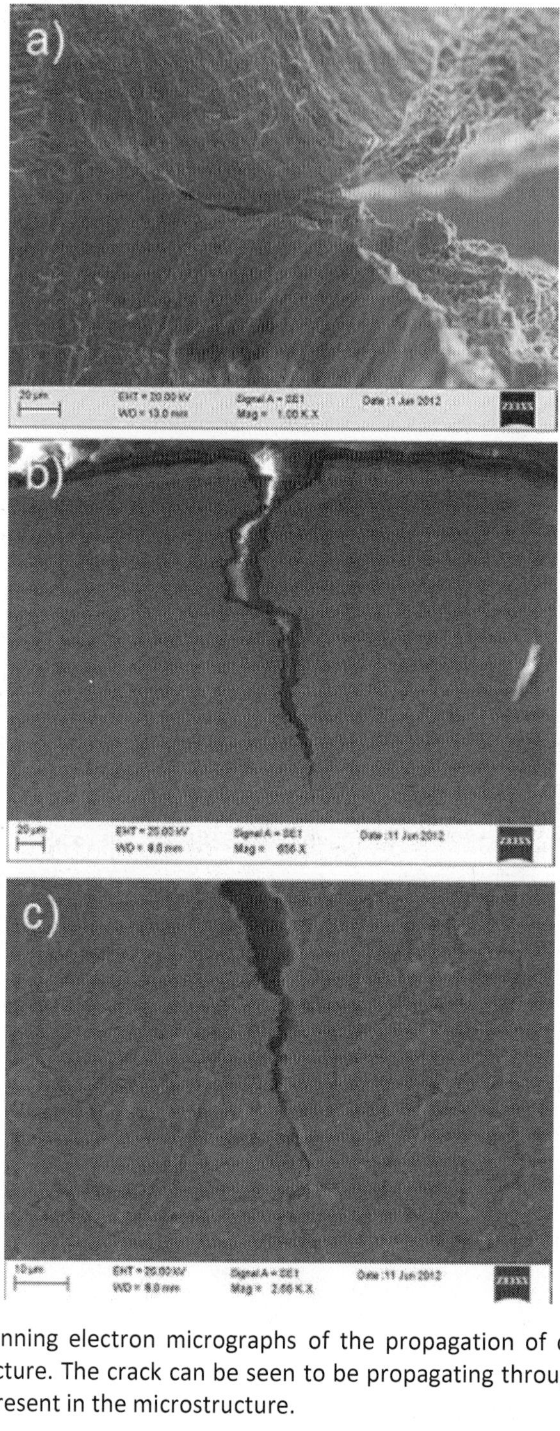

Figure 11. Scanning electron micrographs of the propagation of crack through the microstructure. The crack can be seen to be propagating through the harder beta phases present in the microstructure.

Hardness of the equilibrium phases

The hardness of turrets with α and α+β phases is measured through the micro-hardness measuring instrument. The average values of the hardness of the turret containing α + β phase are 250 ± 20 VPN and that of the α phase is 200 ± 10 VPN. Fig. 12 shows the microhardness indentations taken on turrets with two different microstructures namely α phase (Fig. 12(a)) and α+β phase (Fig. 12(b)).

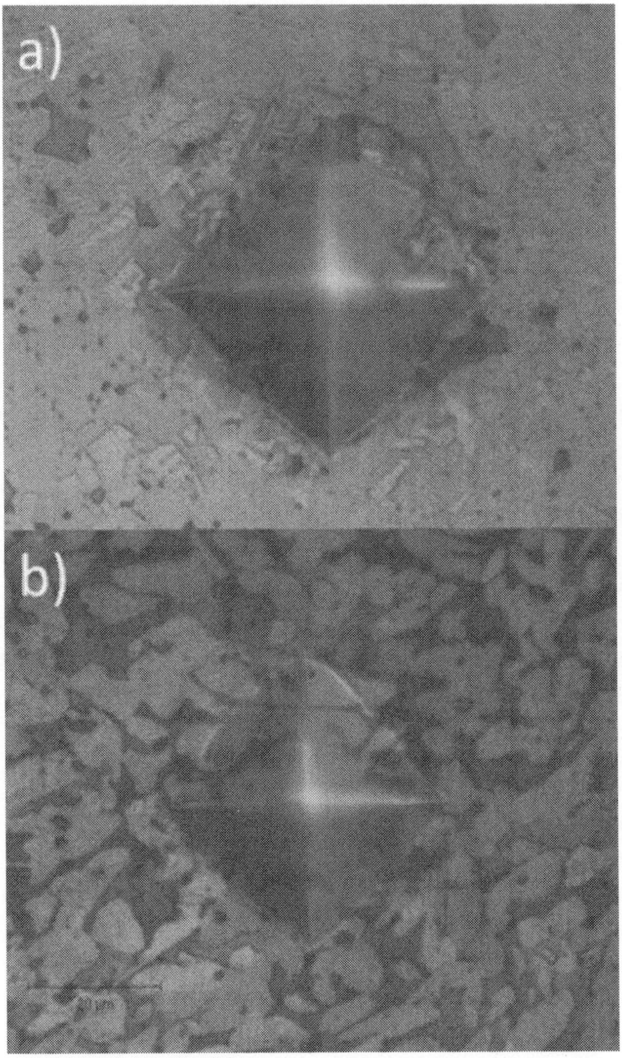

Figure 12. Microhardness indentations taken on (a) turret containing α phase and (b) turret containing α+β phase.

The crack initiates to propagate through the material and the resistance to crack propagation is a function of the toughness of the material. The β phase is harder and so it is more brittle than the α phase. This explains the propagation of the crack through the harder β phase. In free cutting brass with zinc content more than 35%, cracks propagate through the brittle phase in the material. The harder β phase provides a path of least resistance through which the crack is propagated.

In free cutting brass with zinc content less than 35% is a homogeneous α-alpha phase exists. The hardness of α phase is less than β phase and is softer to be cold worked/plastically deformed. The toughness of α phase is hence higher than β-beta phase and hence the material does not rupture during plastic deformation.

CONCLUSION

Free cutting brass as specified by the ASTM B16 has different microstructures for the different composition of the major alloying elements. This affects the material properties, particularly, hardness. The material has homogenous alpha phase at zinc content less than 35 wt.% and the heterogeneous alpha + beta above 35 wt.%. The presence of β-beta phase makes the material more brittle and causes the material to crack during the plastic deformation/cold working. The cracks formed will propagate through the harder β phase in the microstructure. Free cutting brass with zinc content less than 35% has excellent cold workability and hence the specification for the free cutting brass alloy which has to be cold worked/plastically deformed at room temperature can be as follows: copper 62.15–63.0, lead 2.5–3.0, impurities 0.35(max) and zinc – remainder.

The free cutting brass terminals used in the electrical circuits which undergo swaging or cold worked on the printed circuit boards should have single phase composition which will give excellent swaged terminals without cracking. This improves the reliability of the electrical circuit systems

REFERENCES

1. Alloy phase diagrams ASM metals handbook, vol. 3. Metals Park, OH, USA: ASM International; 1992. p. 2.
2. Standard specification for free-cutting brass rod, bar and shapes for use in screw machines ASTM B16, Designation: B 16/B 16M–00, American Society for Testing of Materials, 100 Barr Harbor Drive, West Conshohocken, PA 19428-2959, USA.
3. Balasubramaniam R. Callister's material science and engineering: failure of metals. John Wiley & Sons; 2008. 390.
4. Forming and forging ASM metals handbook, vol. 14. Metals Park, OH, USA: ASM International; 1988. p. 268.

CITATION

Nirmal Ninan, S.V.S. Narayana Murty, Sushant Manwatkar, P. Ramesh Narayanan, Failure analysis of cracked brass turrets used in electronic circuits, Case Studies in Engineering Failure Analysis, Volume 2, Issue 2, October 2014, Pages 107-117, ISSN 2213-2902, http://dx.doi.org/10.1016/j.csefa.2014.04.004.

Chapter 5

A Type of 2D Magnetic Equivalent Circuit Framework of Permanent Magnet for Magnetic System in AEMR

Huimin Liang[1], *Jiaxin You*[1,,], *Zhaowen Cai*[2], *Guofu Zhai*[1]

[1] School of Electrical Engineering and Automation, Harbin Institute of Technology, Harbin 150001, China

[2] Department of Product Design, G&A Technologies Co. Ltd, Guilin 541002, China

ABSTRACT

Modeling of permanent magnet (PM) is very important in the process of electromagnetic system calculation of aerospace electromagnetic relay (AEMR). In traditional analytical calculation, PM is often equivalent to a lumped parameter model of one magnetic resistance and one magnetic potential, but great error is often caused for the inner differences of PM; based on the conception of flux tube, a type of 2D magnetic equivalent circuit framework of permanent magnet model (2D MECF) is established; the element is defined, the relationship between elements is deduced, and solution procedure as well as verification condition of this model is given; by a case study of the electromagnetic system of a certain type of AEMR, the electromagnetic system calculation model is established based on 2D MECF and the attractive force at different rotation angles is calculated; the proposed method is compared with the traditional lumped parameter model and finite element method (FEM); for some types of electromagnetic systems with symmetrical structure, 2D MECF proves to be of acceptable accuracy and high calculation speed which fit the requirement of robust design for AEMR.

INTRODUCTION

Permanent magnet (PM) is widely used in electro-mechanical devices such as aerospace electromagnetic relay (AEMR). The system has a strict reliability requirement with AEMR.[1] And the robust design is an effective way to promote the reliability of AEM.[2]In this process, massive calculation is carried out for the static force or torque calculation of AEMR. There are mainly two methods: finite element method (FEM) and magnetic equivalent circuit method (MEC)[3] FEM has a high-level accuracy but often costs a lot of time. MEC has the superiority in its calculation speed but disadvantages in accuracy. This problem appears more serious in the device which contains permanent magnet (PM).[4] If the accuracy can be improved and remain its fast calculation speed, MEC which is based on analytical method can achieve the robust design mission more suitable.

PM has the characteristics of misalignment between its recoil line and demagnetization curve after serious analysis. Different parts of PM have different magnetic characteristics. That is mainly reflected in that the section magnetic flux of different parts in permanent magnet varies greatly and the working point also varies in different parts. So the equivalence model of PM is quite important and has a great influence on the accuracy of the calculation results.5 [and] 6

Numerous scholars have made studies of the application of inner difference and magnetic circuit equivalent method of PM especially in AEMR. The core purpose is to solve the problem of insufficient accuracy of electromagnetic system analytical calculation caused by PM equivalent model. In the calculation of electromagnetic system which contains PM, it is often regarded as a whole part and it is considered that there is only one working point in PM and it operates on one stable recoil line,7, 8, 9 [and] 10 but actually magnetic property distribution inside the PM is inconsistent.[11] When the magnetic system containing PM is analyzed, MEC has the superiority in its calculation speed but has disadvantages in accuracy compared with FEM.[12] In terms of appearance, leakage permeance in magnetic circuit calculation brings about great error to the calculation results.[13] So multiple studies have been done aimed at various improved calculation methods of leakage permeance.[14] In fact, only basic research directed at PM modeling can fundamentally improve the entire calculation accuracy.[15]

Some scholars referred to the idea of mixing magnetic circuit equivalence to infinitesimal element in the original FEM.[16]They established the solution equation for calculation by extending equivalence from the hexahedron central point to various surfaces. Others adopted similar method to increase the calculation accuracy by the solution procedure of simplified FEM via equivalent magnetic circuit.17 [and] 18 When the electromagnetic system, dimension or the material of permanent magnet changes, artificial division and model have to be re-constructed. PM is the kernel part of a magnetic system and there is no special model for PM. Some work about PM division method which are different from traditional MEC has also been studied.19 [and] 20

In this paper, a type of two-dimension magnetic equivalent circuit framework of permanent magnet model (2D MECF) is established, the element is defined and the solution procedure of this model is analyzed; by a case study of the electromagnetic system of a certain type AEMR, the electromagnetic system calculation model is established based on 2D MECF and the static force at different rotation angles is calculated; traditional lumped parameter model and FEM model are built to verify the validity.

2D MAGNETIC EQUIVALENT CIRCUIT FRAMEWORK

Flux tube

The basic principle of EMC solution comes from the basic principle of electrical circuit. PM is often equivalent to one magnetic resistance and one magnetic potential in EMC method. The magnetic circuit equations for the whole model of electromagnetic system are found. The foundation of EMC is Gauss theorem and the Ampere circuital theorem (Kirchhoff Voltage Law) which is based on Gauss theorem.

In the process of modeling, the parameters of every tiny part are defined by flux tube. The flux tube refers to a flux flow path. The vector which is composed of two end points defines magnetic potential U, and flux ϕ follows the lines which cross the end points (seeFig. 1).

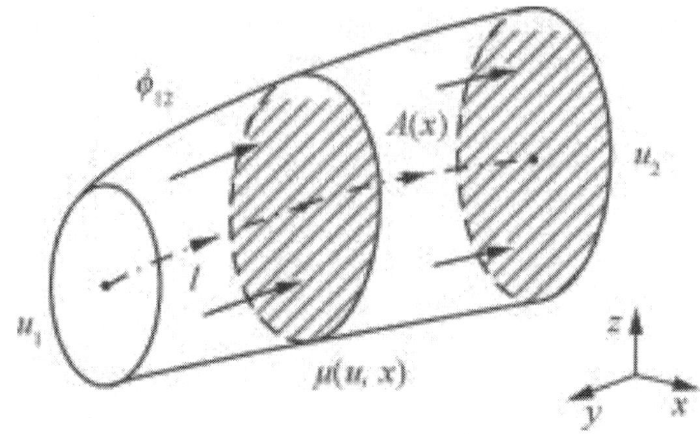

Figure 1. Definition of flux tube.

The permeance of this flux tube is defined as

$$G = \frac{\phi_{12}}{u_1 - u_2} = 1 \Big/ \int_0^l \frac{1}{\mu(u, x) A(x)} \mathrm{d}x \tag{1}$$

where u_1 and u_2 are the magnetomotive forces, ϕ_{12} is the flux through the tube, x the variable of integration alone the flux tube length l, function μ (u, x) the flux tube function (assuming that it is a function associated with magnetic potential and position, and continuous integrable), and $A(x)$ the expression of x–z section plane of flux tube.

In the modeling process, G has different expressing forms. For air and soft magnet, Gcan be solved directly, but for the source such as electromagnetic coil and PM, G can be expressed by outer parameters such as magnetic potential, MMF .etc whose reasonability can be deduced by Eq. (1).

Model of PM potential element

The main reason of great calculation error caused by lumped parameter model is the distribution feature of PM itself. So the suitable way to increase the accuracy is to subdivide the PM. Base on the basic principle of flux tube, the magnetic circuit model can be expressed by a new analytical model.

The element of PM equivalent magnetic framework is shown in Fig. 2. In the element, ϕ_i, ϕ_{i-1} represent the main flux alone the magnetization direction which holds plenty of percentage of flux value; ϕ_{s1}, ϕ_{s2}, ϕ_{s3}, ϕ_{s4} represent the flank flux; *s* represents the section area; U_i, U_{i-1} represent the magnetic potential relative to datum point. For most PM which has regular geometry and normal magnetization way, there is a symmetry axis. The datum point can located at the center of this axis.

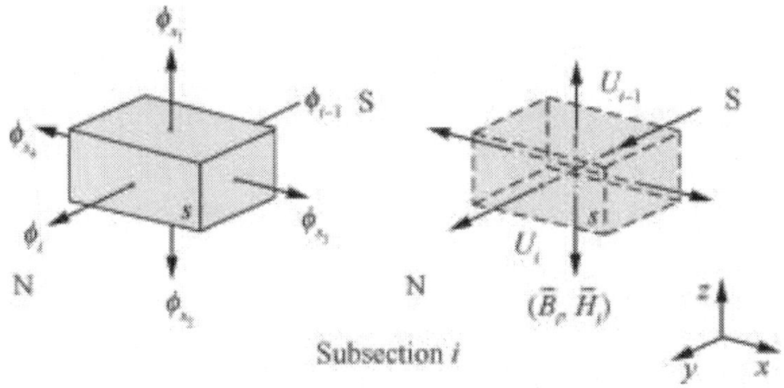

Figure 2. Element of PM equivalent magnetic framework.

For the front and back section, the average section flux can be obtained by $B_i=\phi_i/S$ and $B_{i-1}=\phi_{i-1}/S$ where *S* represents the value of section area. The average magnetic flux density $\overline{B_i}$ can be defined as $\overline{B_i} = (B_i - B_{i-1})/2$ (for enough element division it is obviously right). Correspondingly the average magnetic field intensity $\overline{H_i}$ can be obtained by the feature of PM—demagnetization curve.

For most situations, main flux is predominant. The other sides' flux is relatively less and has little impacts on magnetic attractive force (torque). So the 3D element model can be simplified to 2D framework model which is located in the direction of main flux. For the leakage flux of the third dimension, the calculation error can be compensated by supplement of the leakage permeance.

Relationship between elements and mathematical model

The relationship between elements needs to be defined before the framework is built. The division is carried out in the direction of the place where the most magnetic flux variation occurs—magnetization direction.

For regular shape PM (rectangle cylinder and sector), there is a feature that the magnetic parameters are symmetrical. The relationship between elements is defined based on this feature.

Given the length of the bar nonlinear permanent magnet *L* and section area *A*, the relationship of a pair of two elements is shown in Fig. 3.

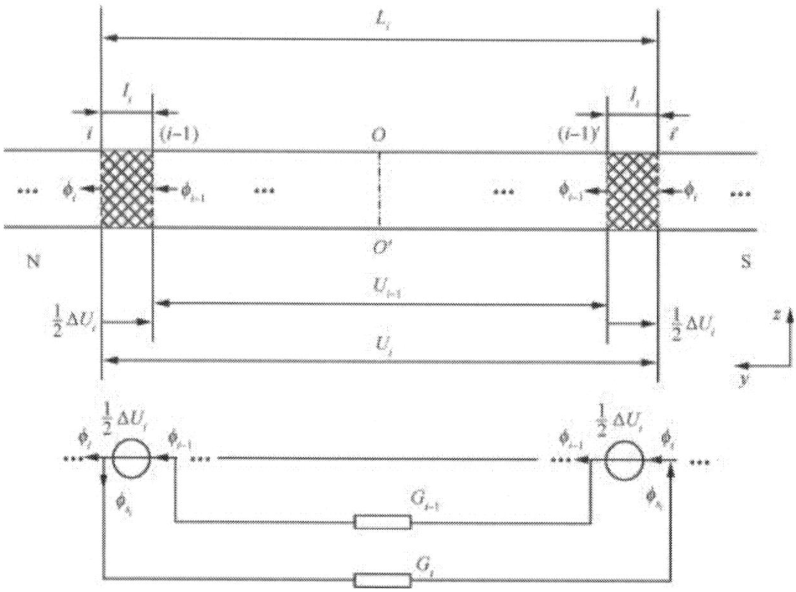

Figure 3. Relationship between a pair of PM elements in 2D MECF.

The PM bar is divided into *n* pairs of elements alone the direction of magnetization where the length of the element is l_i and the distance between the out edge of each pair of element is L_i. For any element pair *i* in Fig. 3, between its two sub-segments which arei-(i-1) and i'–(i-1)', magnetic flux of the inner side surface (i-1)–(i-1)' is ϕ_{i-1}, and magnetic potential is U_{i-1}; magnetic flux of the outer side surface *i*–*i*' is ϕ_i and magnetic potential is U_i. The magnetic potential increment of this segment ΔU_i is

$$\Delta U_i = U_i - U_{i-1} \tag{2}$$

The average magnetic flux $\overline{\phi_i}$ and average magnetic flux density $\overline{B_i}$ of this element pair are

$$\overline{\phi_i} = \frac{1}{2}(\phi_{i-1} + \phi_i) \tag{3}$$

$$\overline{B_i} = \frac{\overline{\phi_i}}{A} = \frac{1}{2}(B_{i-1} + B_i) \tag{4}$$

where $B_i=\phi_i/A$, $B_{i-1}=\phi_{i-1}/A$, *A* represents the value of section area of PM element. According to the PM demagnetization curve $B = f(H)$, the corresponding magnetic field intensity of $\overline{B_i}$ is $\overline{H_i}$, so the magnetic potential increment of element pair *i* is

$$\Delta U_i = -\overline{H_i} l_i \tag{5}$$

Therefore

$$U_{i-1} = -(\overline{H_1} l_1 + \overline{H_2} l_2 + \ldots + \overline{H_{i-1}} l_{i-1}) = -\sum_{\zeta=1}^{j-1} \overline{H_\zeta} l_\zeta \tag{6}$$

Considering Eqs. (2), (5) and (6), average magnetic potential $\overline{U_i}$ of element *i* has two equal expressions

$$\overline{U_i} = \frac{1}{2}(U_{i-1} + U_i) = \begin{cases} U_{i-1} + \frac{1}{2}\Delta U_i = -(\sum_{\zeta=1}^{i-1} \overline{H_\zeta} l_\zeta + \frac{1}{2}\overline{H_i} l_i) \\ U_i - \frac{1}{2}\Delta U_i = -(\sum_{\zeta=1}^{i} \overline{H_\zeta} l_\zeta - \frac{1}{2}\overline{H_i} l_i) \end{cases} \tag{7}$$

The leakage permeance of the element is

$$dG_{si}=\mu_0 g_{si}(x)dx \tag{8}$$

where μ_0 is the permeability of air, $g_{si}(x)$ is the relative leakage permeance. According to Eq. (6), magnetic leakage flux ϕ_{si} is

$$\phi_{s_i} = \int_{\frac{l}{2}} d\phi_{s_i} = \frac{1}{2}\mu_0 \int_0^{\frac{l}{2}} \bar{U}_i g_{s_i}(x)dx = \begin{cases} -G_i(\sum_{\zeta=1}^{i-1} \overline{H_\zeta} l_\zeta + \frac{1}{2}\overline{H_i} l_i) \\ -G_i(\sum_{\zeta=1}^{i} \overline{H_\zeta} l_\zeta - \frac{1}{2}\overline{H_i} l_i) \end{cases} \tag{9}$$

For $\phi_{i-1}=\phi_i+\phi_{si}$, combined with Eqs. (3), (4) and (9), it can be deduced that

$$\overline{B_i} - B_i = B_{i-1} - \overline{B_i} = -\frac{G_i}{2A}\left(\sum_{\zeta=1}^{i-1}\overline{H_\zeta} l_\zeta + \frac{1}{2}\overline{H_i} l_i\right) \tag{10}$$

For the first pair, i = 1, it can be gained according to Eq. (8) that

$$B_1 = \overline{B_1} + \frac{G_1}{4A}\overline{H_1} l_1 \tag{11}$$

Thus, the entire mathematical model of element pairs is summarized in Eq. (12).

In the practical modeling process, the magnetic field line division method is used especially for an open circuit PM:

$$\begin{cases} B_1 = \overline{B_1} + \dfrac{G_1}{4A}\overline{H_1} l_1 \\ B_{i-1} - \overline{B_i} = -\dfrac{G_i}{2A}\left(\displaystyle\sum_{\zeta=1}^{i-1}\overline{H_\zeta} l_\zeta + \frac{1}{2}\overline{H_i} l_i\right) \quad (i = 2, 3, \ldots, n) \\ B_i = 2\overline{B_i} - B_{i-1} \\ \overline{B_i} = f(\overline{H_i}) \end{cases} \tag{12}$$

End verification condition

The whole solution process is determined by initial value and the correctness is tested by end verification condition. The actual end verification condition is related to the actual magnetic system. End verification condition of a series of element pairs of PM model is

$$B'_n = \frac{G_{\text{end}}\overline{U_n}}{A} = \frac{G_{\text{end}}}{A}(\overline{H_1} l_1 + \overline{H_2} l_2 + \ldots + \overline{H_n} l_n) \tag{13}$$

where G_{end} represents the end leakage permeance which can be calculated by the equivalent spherical pole calculation theory. 21 [and] 22

APPLICATION OF 2D MECF IN ELECTROMAGNETIC SYSTEM OF AEMR

Electromagnetic system of AEMR

The 2D MECF method is applied to a certain type of AEMR to build the model of magnetic system and calculate the attractive force.

The experimental model of an AEMR and its structure are shown in Fig. 4 (the coils on the two iron cores have not been assembled).The permanent magnet is rectangle bar whose magnetization direction is alone the length. The yoke pillar and the base are made of soft magnet material.

(a) Experimental model of an AEMR

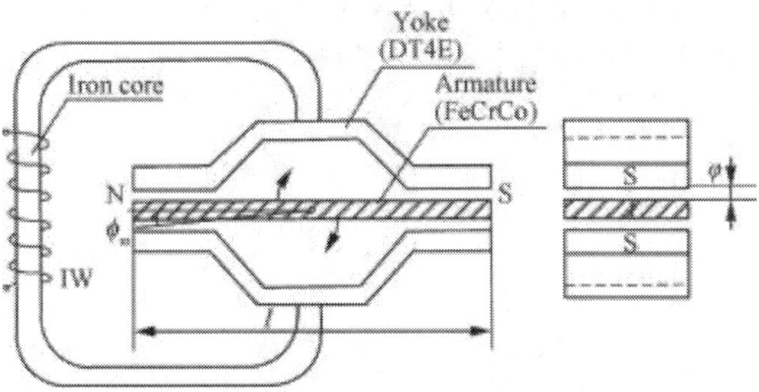

(b) Structure diagram of the AEMR

Figure 4. Electromagnetic structure of a certain type of AEMR.

This type of AEMR has the feature of symmetry and its structure parameters are shown in Table 1.

Table 1. Configuration parameters of AEMR.

Symbol	Quantity	Value
l	Length of armature	20 mm
φ	Air gap	1.0 mm
S_a	Section area of armature	5.94 mm^2
S_y	Section area of yoke	8.1 mm^2
θ_m	Maximum rotation angle	5.48°

According to the distribution feature of PM magnetic field, the whole 2D plane can be divided into four regions:

1) Region I. Center of the PM bar (magnetic field lines are rare).
2) Region II. Middle of the PM sides (magnetic field lines are little and flow through the yoke and PM).
3) Region III. Terminal of the PM sides (magnetic field lines are intensive and flow through the yoke, thin gap and PM).
4) Region IV. Outer area. Each region can be divided into any number of elements according to specific calculation requirement. It can be found that this division method is reasonable especially when investigating the distribution of 2D FEM magnetic field lines. The region division sketch map is shown in Fig. 5.

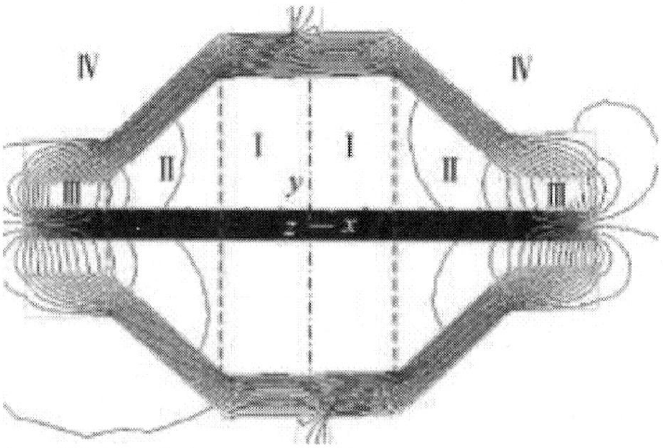

Figure 5. Region division of electromagnetic system.

The demagnetization curve of PM and the magnetization curve of soft magnet are shown in Fig. 6.

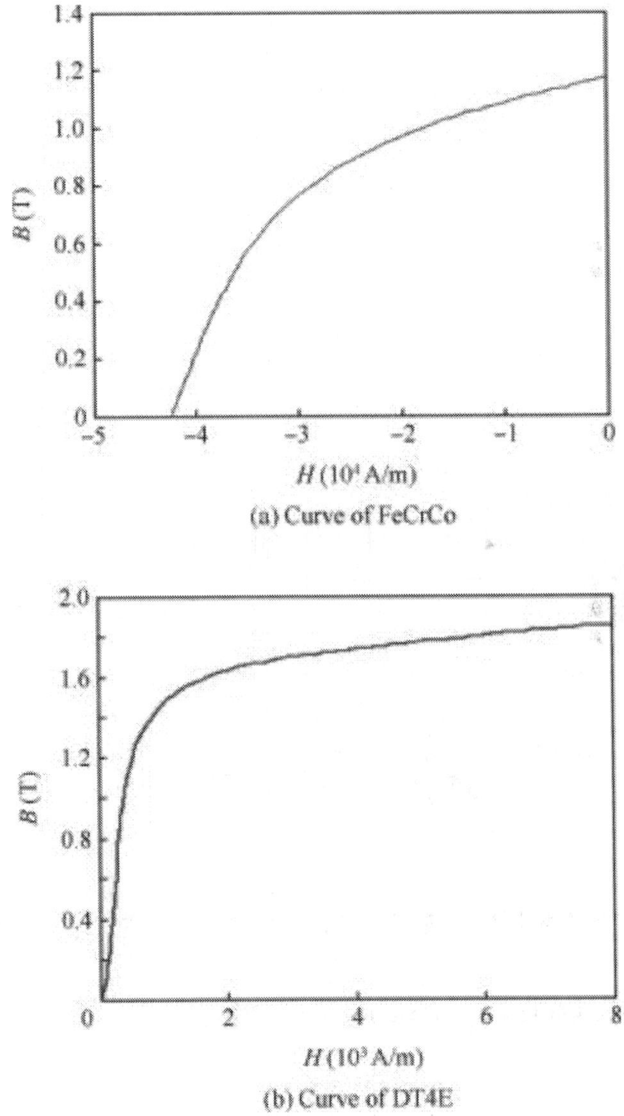

Figure 6. Demagnetization curve of PM and magnetization curve of soft magnet.

Building of 2D MECF

Region I and II will be divided equally and Region III will be fine divided into elements. According to the flux tube closed way, elements will be set for two types: leakage flux flows via the soft magnet and leakage flux flows via only the air.

The division is shown in Fig. 7. The leakage flux of element pair 1-1′ closes via the air and element pair (2-2′)–(6-6′) closes via soft magnet.

(1)For element pair 1–1′, the flux closes from PM to air and back to PM. According to the model shown in Section 2.3, the equations are summarized as

$$\begin{cases} \Delta\phi_i = \phi_{i-1} - \phi_i \\ \Delta\phi_i = G_i\overline{U_i} \\ \overline{U_i} = \overline{H_i}l_i \\ B_i - B_{i-1} = \Delta\phi_i/S_a \\ \overline{B_i} = (B_i + B_{i-1})/2 \end{cases} \tag{14}$$

where B_i, $\overline{B_i}$ and H_i, $\overline{H_i}$ satisfy the demagnetization function B=f(H) .

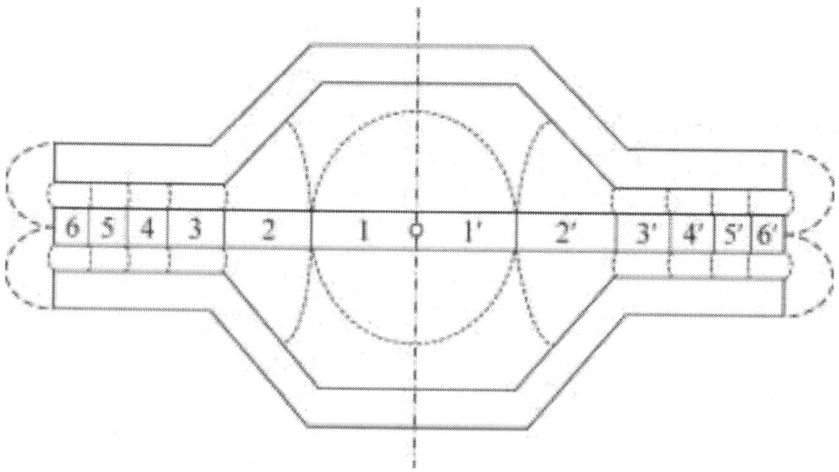

Figure 7. Division of PM in magnetic system.

From Eq. (14) it can be deduced that

$$B_i S_a - B_{i-1} S_a = G_i \overline{H_i} l_i \tag{15}$$

So there is

$$(B_i - B_{i-1}) S_a = G_i f(B_i/2 + B_{i-1}/2) l_i \tag{16}$$

where section S_a and the leakage permeance G_i are known, an initial value of B_1 can solve a B'_n, and the end condition will verify the value of B'_n; if $|B_n - B'_n| > \varepsilon$ (ε is set as error value), B_1 should be reassigned and solution process repeats till $|B_n - B'_n| \leqslant \varepsilon$.

Leakage permeance is obtained using analytical method. The leakage permeances use the sum form which considers the leakage flux of the third dimension. The formula in this magnetic system is

$$G_i = \sum G_{i(x,y)} + \sum G_{iside} \tag{17}$$

$$G_{i(x,y)} = \mu_0 \int_x \int_y \frac{\mathrm{d}S_{s_i}(x,y)}{l_{s_i}(x,y)} \tag{18}$$

where G_i represents leakage permeance of element pair *i* , $G_{i(x,y)}$ the leakage permeance in *x-y* plane, G_{iside} the inclined-side leakage permeance, $S_{si}(x,y)$ the equivalent area of flux tube and $l_{si}(x,y)$ the equivalent length of flux tube.

(2)For element pair (2–2′)–(6–6′), the flux closes from PM to soft magnet and to PM. The framework of model is shown in Fig. 8.

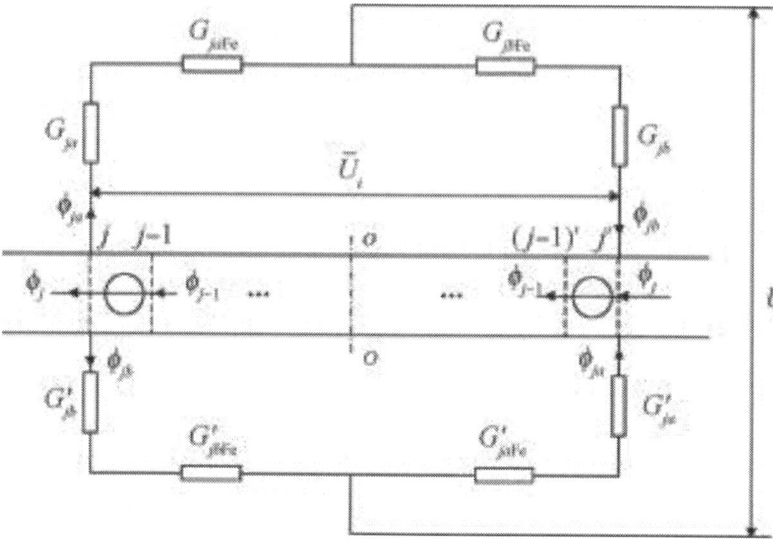

Figure 8. MECF model of element pair (2-2′)–(6-6′).

In Fig. 8G*ja* and G*jb* represent the air leakage permeance of each element pair (when armature rotates, permeance of air gap, flux of air gap, permeance of soft magnet and the end leakage permeance can be divided into up and down parts, distinguished by the "′", in which j = 2, 3, 4, 5, 6); G_{jaFe}, G_{jbFe} represent soft magnet permeance; in this framework $G_{jaFe} = G'_{jaFe}$, $G_{jbFe} = G'_{jbFe}$, $G_{ja} = G'_{ja}$ and $G_{jb} = G'_{jb}$; U_l r epresents the magnetic potential of the coil.

The equations are summarized as

$$\begin{cases} \Delta\phi_j = \phi_{j-1} - \phi_j \\ \Delta\phi_j = \phi_{ja} + \phi_{jb} \\ \overline{U_j} = \overline{H_j} l_j \\ B_j - B_{j-1} = \Delta\phi_j / S_a \\ \overline{B_j} = (B_j + B_{j-1})/2 \end{cases} \tag{19}$$

The main difference is the change of $\Delta\phi_j$ compared with Eq. (14). The average magnetic potential is

$$\overline{U_j} = \phi_{ja}/G_{ja} + \phi_{ja}/G_{jaFe} + \phi_{jb}/G_{jb} + \phi_{jb}/G_{jbFe} \tag{20}$$

For uniform material, it can be deduced that

$$\begin{cases} \phi_{ja} = G_{ja\mathrm{Fe}}\overline{U}_{ja\mathrm{Fe}} \\ \overline{U}_{ja\mathrm{Fe}} = \overline{H}_{ja\mathrm{Fe}} l_{ja\mathrm{Fe}} \\ \overline{B}_{ja\mathrm{Fe}} = f_{\mathrm{Fe}}(\overline{H}_{ja\mathrm{Fe}}) \\ \phi_{ja} = \overline{B}_{ja\mathrm{Fe}} S_y \end{cases} \tag{21}$$

where f_{Fe} is the magnetization curve formula of soft magnet. For Gjb_{Fe} there is also a similar formula to Eq. (21).

When $IW = 0$ the magnetic potential is nearly zero, that is $U_l = 0$. Simulcast Eqs. (19),(20) and (21), the solution formula can be deduced:

$$\begin{aligned} &\Delta\phi/G_{ja} + l_{ja\mathrm{Fe}} f_{\mathrm{Fe}}^{-1}(\Delta\phi/S_y) + \\ &\Delta\phi'/G_{jb} + l_{jb\mathrm{Fe}} f_{\mathrm{Fe}}^{-1}(\Delta\phi'/S_y) + l_j f^{-1}(\sum\phi/S_y) = 0 \end{aligned} \tag{22}$$

where $\Delta\phi=(B_{j-1}S_y-B_jS_y)/(1+Gja/Gjb)$ $\Delta\phi'=(B_{j-1}S_y-B_jS_y)/(1+Gjb/Gja)$ $\sum\phi=(B_{j-1}S_y+B_jS_y)/2$. G_{ja} and Gjb can be obtained by analytical method. l_j, l_{jaFe}, l_{jbFe} and S_y are known. So the parameter B_j can be solved.

(3)MECF of magnetic system

Based on the analysis above, the whole 2D MECF model of the magnetic framework model of AEMR is shown in Fig. 9.

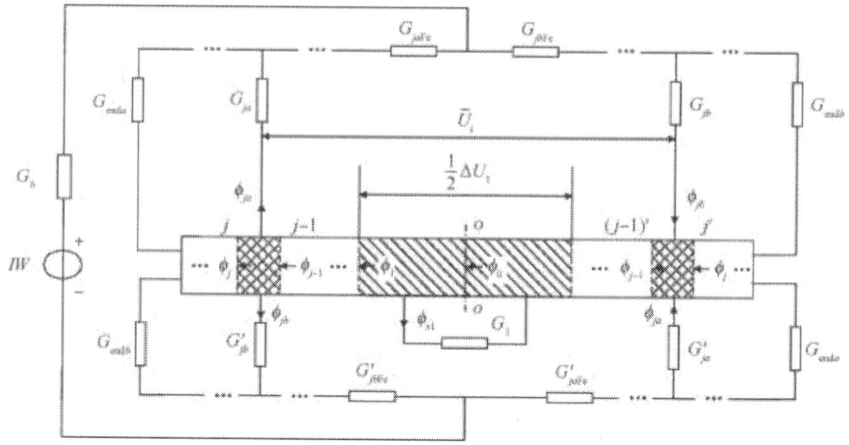

Figure 9. AEMR electromagnetic system MECF model.

Solving of total torque

The force of each element pair of PM armature can be calculated by Maxwell electromagnetic force formulas. Adding with each element's torque, the attractive force of PM armature can be calculated. There is an assumption that the magnetic flux density of each subsection surface is uniform (especially under large subsection number). After the iteration of magnetic flux density solving, the attractive force can be calculated using the formulas below:

$$\begin{cases} F_{1i} = (\Delta\phi_{1i})^2/(2\mu_0 S_{1i}) \\ F_{2i} = (\Delta\phi_{2i})^2/(2\mu_0 S_{2i}) \end{cases} \tag{23}$$

where S_{1i} and S_{2i} represent the area of each subsection that flux flows through.

The force calculation sketch map is shown in Fig. 10.

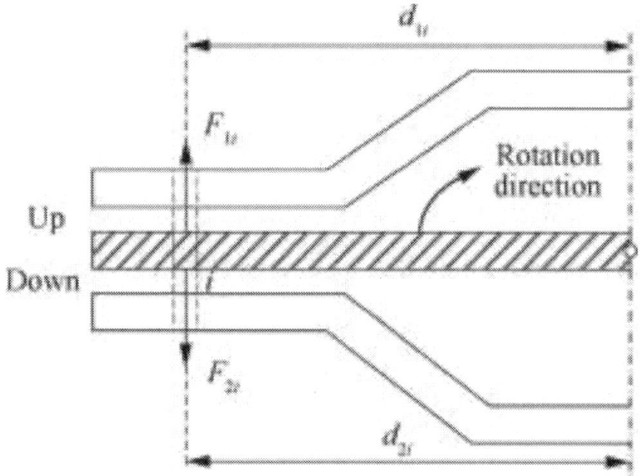

Figure 10. Sketch map of force calculation.

For the symmetrical air gap in bridge magnetic system mentioned above, the total torque of one side $M_{oneside}$ can be calculated as

$$M_{oneside} = \sum_{i=1}^{n} 2(F_{1i}d_{1i} - F_{2i}d_{2i}) \tag{24}$$

where d_{1i} and d_{2i} represent the torques to the center of F_{1i} and F_{2i} . When the rotation angle is not equal to 0, there is $d_{1i} \neq d_{2i}$.

Comparison and analysis of the results

The range of the rotation angle of this electromagnetic system is −5.48° to 5.48° For its symmetrical structure, rotation range of 0–5.48° need to be calculated.

To compare with the results of 2D MECF, MEC model of this AEMRis also built. The magnetic circuit is shown in Fig. 11 where R_1 and R_2 represent the magnetic resistance of armature air gap; R_1' and R_2' represent the magnetic resistance of armature air gap (opposite side); F_{PM} represents the equivalent magnetic potential of PM armature, $R_{PM}$the equivalent magnetic resistance of PM armature, and R_{SB} the resistance of the base;R_{SR} and R_{SL} represent the magnetic resistance of left and right sides of soft magnet material;.

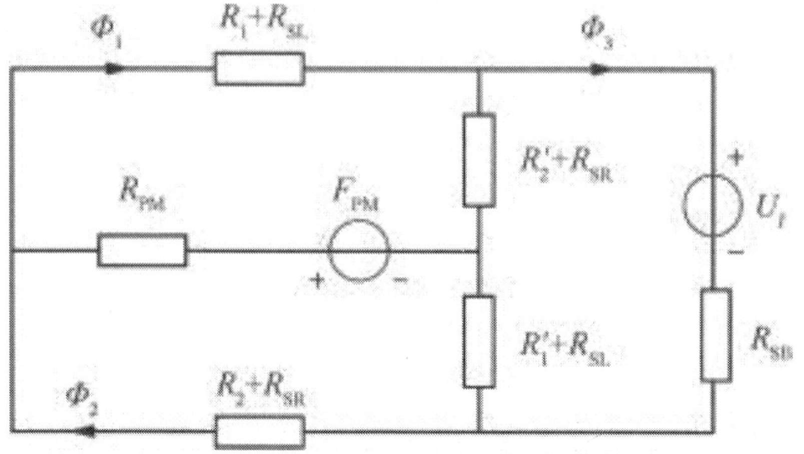

Figure 11. MEC model of AEMR.

FEM model is also built to verify the calculation results. The 3D FEM and 2D FEM magnetic flux intensity distribution sat 5.4° are shown in Fig. 12. The number of 3D FEM model elements is 22173.

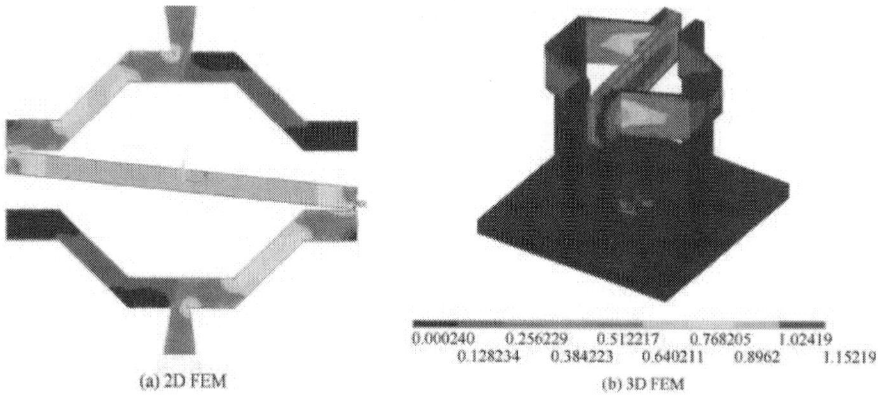

Figure 12. FEM magnetic flux intensity distribution.

On the software platform of Matlab, the attractive force of the 2D MECF model of AEMR shown in Fig. 9 is calculated under zero ampere-turn. The method of Newton–Raphson is adopted to solve the parameters of 2D MECF model. Fig. 13 shows the 3D distribution of magnetic field intensity of each element at different angles when $IW = 0$.

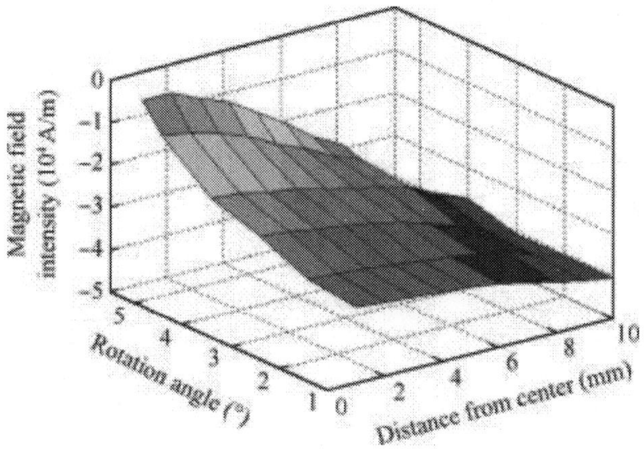

Figure 13. 3D distribution of magnetic field intensity of each element at different angles *IW* = 0.

With the increase of rotation angle, the value of magnetic field intensity from center to end of PM is getting small. The graph of model results varies smoothly in Fig. 13.

The attractive force calculation results and the error contrast 3D FEM are shown in Table 2.

Table 2. Calculation results of 2D MECF, MEC and FEM.

Rotation angle (°)	MEC (N)	2D FEM (N)	2D MECF (N)	3D FEM (N)
1	0.0917	0.0316	0.0296	0.0324
2	0.2168	0.0925	0.0753	0.0819
3	0.3676	0.1901	0.1403	0.1577
4	0.6198	0.3094	0.2627	0.2782
4.5	0.8321	0.4721	0.4054	0.3788
5	1.1385	0.703	0.6112	0.5391
5.4	1.4746	1.0279	0.7577	0.7413
Max error	165%	38.70%	13.40%	
Average error	133.30%	19.50%	7.99%	

The MEC, 2D MECF, 2D FEM and 3D FEM calculation results are shown in Fig. 14.

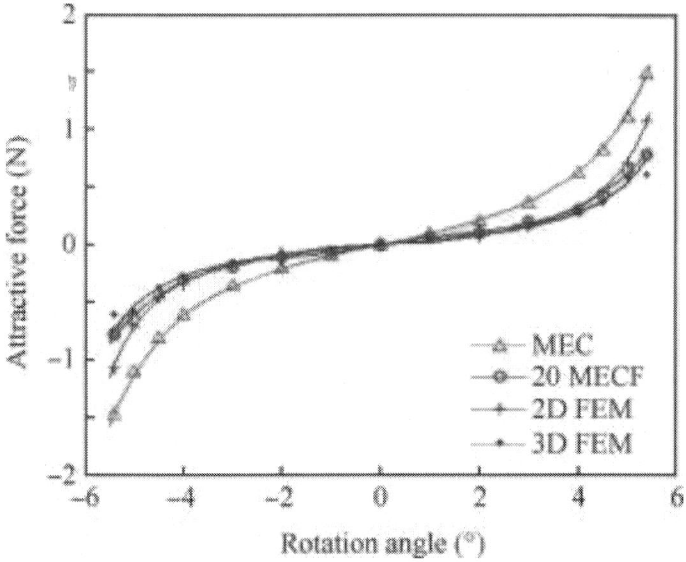

Figure 14. Comparison of MEC, 2D MECF and FEM calculation results.

It can be seen from Fig. 14 that compared with lumped parameter equivalent magnetic circuit method and 2D FEM, the calculation results of 2D MECF approximate the calculation results of 3D FEM to a larger extent. Compared with 3D FEM the maximum error is 13.4% and the average calculation error of 2D MECF is 7.99%.

The fitting curve is given according to the FEM results and 2D MECF results. From Fig. 14 it can be drawn that the 2D MECF results approximate the calculation results of 3D FEM and the overall trend are similar. The reason for error is analyzed:

(1)The end leakage permeance of 2D MECF when rotation angle is big is not accurate enough, which leads to the error of force calculation.

(2)The number of elements in the model above is small and more subsections will bring upgrade of the accuracy.

For the ampere-turns of $IW = \pm 120$, the results of 2D MECF and 3D FEM model are compared in Fig. 15.

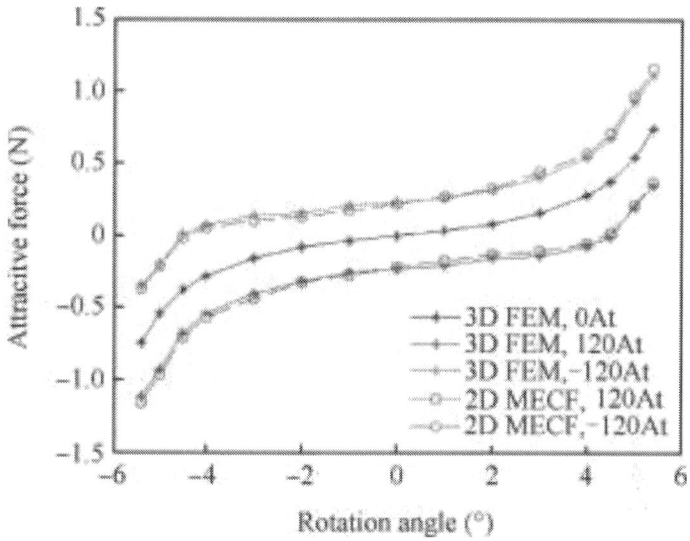

Figure 15. Comparison of 2D MECF and FEM model under ± 120At.

The maximum error is 15.9% andthe average error of 2D MECF is 8.65% compared with FEM model under the action ampere-turns.

About the calculation time, the situation of *IW* = 0 is taken for example. In the same PC hardware and Matlab programming environment, it takes less than 1 s for calculation via traditional equivalent magnetic circuit model, less than 5 s for calculation via 2D MECF model, 56 s for calculation via 2D FEM (number of element: 9877), and 1200 s for calculation via 3D FEM (number of element: 22173), and the time required by 2D MECF model is far less than 3D FEM method.

CONCLUSIONS

1)According to the fundamental principle of flux tube, the definition of 2D MECF element has been presented; the parameters of this element have been defined, the relationship between pairs of elements have been provided, the whole mathematical model has been summarized and the solution procedure of this model has been analyzed.

2)The 2D MECF has been adopted to a certain type of AEMR which contains PM. The PM element region has been divided; the 2D MECF model of the magnetic system has been built; the parameters of PM elements at different angles and different ampere-turns have been solved; the attractive force of the armature has been calculated; compared with 3D FEM when *IW* = 0 the maximum error is 13.4% and the average error of 2D MECF is 7.99%; when *IW* = ±120 the maximum error is 15.9% and the average error of 2D MECF is 8.65%.

3)Compared with the 3D FEM, 2D FEM and MEC results, 2D MECF results are consistent with 3D FEM results to a large extent, and the time cost is much less than 3D FEM. The characteristics of high calculation speed and moderate accuracy of 2D MECF can provide different thoughts for rapid modeling and parameters' design process for electromechanical devices which belong to some type of symmetrical structure AEMR and contain PM.

ACKNOWLEDGEMENTS

The authors express their thanks to the support of National Natural Science Foundation of China (Nos. 51177021, 51207028).

REFERENCES

1. Lee SJ, Kim KS, Cho S, Jang JY, Lee TH, Hong JP. Optimal design of interior permanent magnet synchronous motor considering the manufacturing tolerances using taguchi robust design. Electr Power Appl 2014;8(1):23–8.
2. Ye XR, Deng J, Ma Y, Zhai GF, Yan JH. A robust solution for hesitate phenomenon in pick up process of aerospace electromagnetic relay. Chin J Aeronaut 2013;26(3):697–704.
3. Liang HM, Lin JB, Zhai GF. Method for identification of nonlinear parameters and its application to data analysis for aerospace relay reliability. IEICE Trans Electron 2006;89(8): 1173–6.
4. Ives L, Bui T, Vogler W, Bauer A, Shephard M, Beall M. Threedimensional magnetostatic analysis using structured mesh and nodal elements. IEEE Trans Magn 2011;47(2):198–202.
5. Gorazd G, Gorazd L, Damijan M. Analytical model of permeance variation losses in permanent magnets of the multipole synchronous machine. IEEE Trans Magn 2013;49(2):921–8.
6. Christian SW, Polinder H. Analytical-numerical hybrid model for flux-switching permanent magnet machines. Proceeding of the 39th annual conference of the IEEE Industrial Electronics Society; 2013. p. 2614–9.
7. Hsieh Mu, Lin IH, Hsu YC, McMahon, Richard A. Design of brushless doublyfed machines based: on magnetic circuit modeling. IEEE Trans Magn 2012;48(11):3017–20.

8. Sharma Vijay, Kumar GV, Dash SK, Nashine BK, Rajan KK. Modeling of permanent magnet flowmeter for voltage signal estimation and its experimental verification. Flow Meas Instrum 2012;28(12):22–7.
9. Wu S, Jiao ZX, Yan L, Yu JT, Chen CY. A fault-tolerant tripleredundant voice coil motor for direct drive valves: design, optimization, and experiment. Chin J Aeronaut 2013;26(4):1071–9.
10. Mouton Z, Kamper MJ. Modeling and optimal design of an eddy current coupling for slip-synchronous permanent magnet wind generators. IEEE Trans Industr Electron 2014;61(7):3367–76.
11. Manz B, Benecke M, Volke F. A simple, small and low cost permanent magnet design to produce homogeneous magnetic fields. J Magn Reson 2008;192(1):131–8.
12. You JX, Liang HM, Ye XR, Zhai GF. The effect of coil on combined three-subsection permanent magnet in close magnetic circuit modelProceeding of the 57th IEEE holm conference on electrical contacts; 2011 Sep 13–15. Minneapolis, US: IEEE, Inc; 2011. p. 175–8.
13. Wallin M, Ranlo¨f M, Lundin U. Reduction of unbalanced magnetic pull in synchronous machines due to parallel circuits. IEEE Trans Magn 2011;47(12):4827–33.
14. Rafael EP, Enrique M, Jose AR. Systematic coupling of multiple magnetic field systems and circuits using finite element and modified nodal analyses. IEEE Trans Magn 2011;47(1): 207–13.
15. Bi LX, Wang ShM, Xia YH. Calculation of leakage permeance in a surface-mounted permanent magnet machine. J Tsinghua Univ (Sci Technol) 2010;50(4):525–8.
16. Zhai GF, Wang QY, Ren WB. An output space-mapping algorithm to optimize the dimensional parameter of electromagnetic relay. IEEE Trans Magn 2011;47(9):2194–9.
17. Amrhein M, Krein PT. 3D magnetic equivalent circuit framework for modeling electromechanical devices. IEEE Trans Energy Convers 2009;24(2):397–405.
18. Amrhein M, Krein PT. Force calculation in 3D magnetic equivalent circuit networks with a maxwell stress tensor. IEEE Trans Energy Convers 2009;24(3):587–93.
19. Liang HM, You JX, Yang WY, Zhai GF. A type of subsection model for a permanent magnet bar and its leakage permeance calculation method in an open magnetic circuit. J Magn 2014;19(1):37–42.
20. Tiegna H, Amara Y, Barakat G. Overview of analytical models of permanent magnet electrical machines for analysis and design purposes. Math Comput Simul 2013;90:162–77.
21. Zhang GS. Basic theory of electrical apparatus. Beijing, China: Machine Press; 1989. p. 208–15.

22. Liang HM, You JX, Luo FB, Yang WY, Zhai GF. A distributed parameter model of open circuit nonlinear permanent magnet bars based on magnetic field lines. Proc Chin Soc Electr Eng 2014;34(9):1429–35.

CITATION

Huimin Liang, Jiaxin You, Zhaowen Cai, Guofu Zhai, A type of 2D magnetic equivalent circuit framework of permanent magnet for magnetic system in AEMR, Chinese Journal of Aeronautics, Volume 28, Issue 1, February 2015, Pages 268-276, ISSN 1000-9361, http://dx.doi.org/10.1016/j.cja.2014.12.021.

Chapter 6

Performance Prospects of Fully-Depleted SOI MOSFET-Based Diodes Applied To SCHENKEL Circuit for RF-ID Chips

Yasuhisa Omura, Yukio Iida

Department of Electric, Electronics and Informatics, Kansai University, Suita, Japan

ABSTRACT

The feasibility of using the SOI-MOSFET as a quasi-diode to replace the Schottky-barrier diode in the Schenkel circuit is examined by device simulations primarily and experiments partly. Practical expressions of boost-up efficiency for d. c. condition and a. c. condition are proposed and are examined by simulations. It is shown that the SOI-MOSFET-based quasi-diode is a promising device for the Schenkel circuit because high boost-up efficiency can be gained easily. An a. c. analysis indicates that the fully-depleted condition should hold to suppress the floating-body effect for GHz-level RF applications of a quasi-diode.

INTRODUCTION

Since RF-ID chips have no internal power supply, they need a way of using the received signal as an energy source; a common approach is the Schenkel circuit [1,2]. The basic Schenkel circuit is shown in **Figure 1**. It usually consists of capacitors and pn diodes (PND's) or Schottky-barrier diodes (SBD's). Modern RF applications such as RF-ID chips often use

Schottky-barrier diode (SBD) in this circuit [2-4]. Unfortunately, generally speaking, the reverse-biased current (I_R) of an SBD is not significantly lower than the forward-biased current (I_F) because the requirement for high drive currents results in a low barrier height. The reverse-biased current should be extremely low because the a. c. signal voltage received is very small in RF-ID systems. Overall, generally speaking, Schenkel circuits that use SBD's fail to offer high boost-up efficiency, resulting in many stages of boost-up circuit block.

Recently, RF-ID chips are applied to various systems without limitation of production costs because applications to social security including living safe attract attention [5]. In these cases, performance and reliability are primarily important. So, a new market of RF-ID chips is growing up.

On the other hand, SOI MOSFET is one of promising devices that can be applied to RF circuit applications [6]

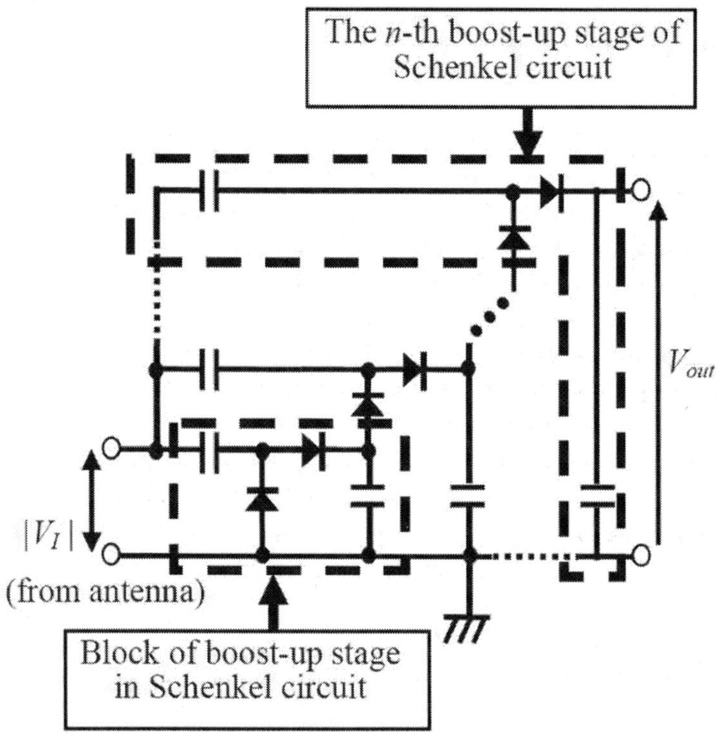

Figure 1. Schematic of Schenkel circuit.

because high-resistivity substrate can be easily introduced [7]. Since the high-resistivity substrate presents not only a low-loss transmission of RF signal [7], but also a low cross-talk in digital circuits [8], various applications are already reported [9,10].

In this paper, we discuss using the SOI-MOSFETbased quasi-diode (SOI-QD) to replace that SBD in Schenkel circuits. First we propose the expression of boost-up efficiency for a low-frequency range using experimental d. c. characteristics of SOI-QD made from various SOI MOSFET's, and a. c. analyses of SOI-QD are conducted using a two-dimensional (2D) device simulator (ISE DESISS [11]) to investigate operation stability in the RF band. We also define another expression of boost-up efficiency in the RF band, and examined its availability on the basis of a. c. simulation results. RF-band potential of SOI-MOSFET-based quasi-diode is addressed from the viewpoint of future RF applications.

REMAINING ISSUES OF CONVENTIONAL SCHENKEL CIRCUIT AND AN ADVANCED PROPOSAL

At first, we used the circuit simulator PSPICE [12,13] to examine the performance of a Schenkel circuit that used Schottky barrier diode (SBD), pn-junction diode (PND) or conventional bulk MOSFET-based quasi-diode (CBQD); in the CB-QD variant, the gate terminal and the drain terminal are connected and the source terminal and the substrate terminal are connected. We assumed that the SBD and PND had a junction area of 46.1 μm^2, and that the gate width and the gate length of the bulk MOSFET were 20.6 μm and 0.32 μm. All devices had identical active areas; the junction area of bulk MOSFET is 39.5 μm^2. The circuit simulations employed the empirical model (Level = 3) for simplicity [12]. To acquire realistic device performance from the PND and SBD variants, we introduced the minority carrier lifetime model shown in Appendix A.

Figure 2 shows simulated rectifier characteristics of the various diodes in a low voltage range of input anode voltage (V_A). We can see that SBD with the barrier height (f_b) of 0.15 eV has the largest driving current among the three diodes, but it has the highest reverse-biased current (4.43 μA). **Figure 3** shows the performance of 5-stage Schenkel circuits that use the three different diodes for an input voltage $|V_I|$ of 100 mV. It is shown that the conventional bulk-MOSFET-based quasi-diode (CB-QD) successfully boosts the input signal from a very low level to an acceptable level, while SBD and pn-junction diode fail to do so. SBD failed to match this despite its large driving current; since current SBD designs have a high I_R value, almost identical to I_F, in the low voltage range of 100 mV, the high leakage current (I_R) degrades the signal boost process. On the other hand, the I_F and I_R values of CB-QD are much smaller than those of SBD. However, CB-QD offers an acceptable level of boost. The main reason is that I_F of CB-QD is larger than own I_R, which means that the effective boost efficiency (η) of a Schenkel circuit should not be determined by the direct value of driving current, but by the ratio of I_F to I_R defined as

$$\eta = I_F / \left(I_F + |I_R|\right). \qquad (1)$$

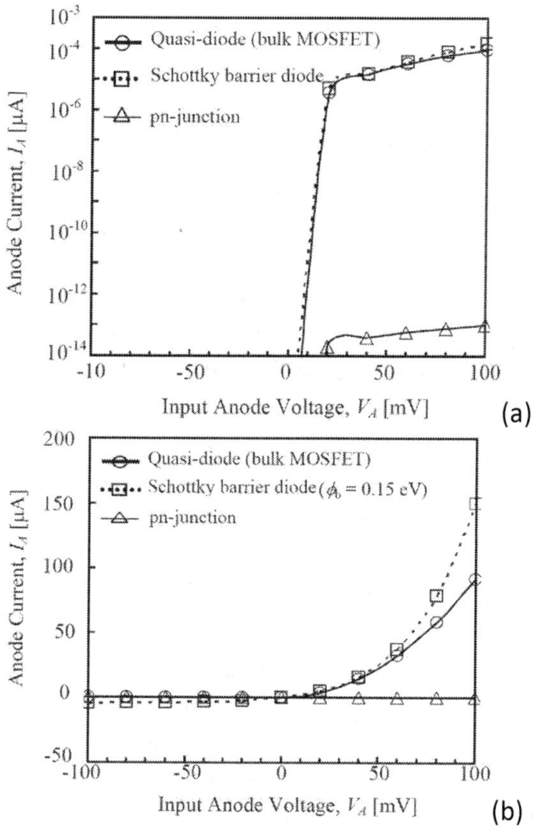

Figure 2. Forward and reverse characteristics of various diodes (Pspice simulation results). (a) Log scale plot; (b) Linear scale plot.

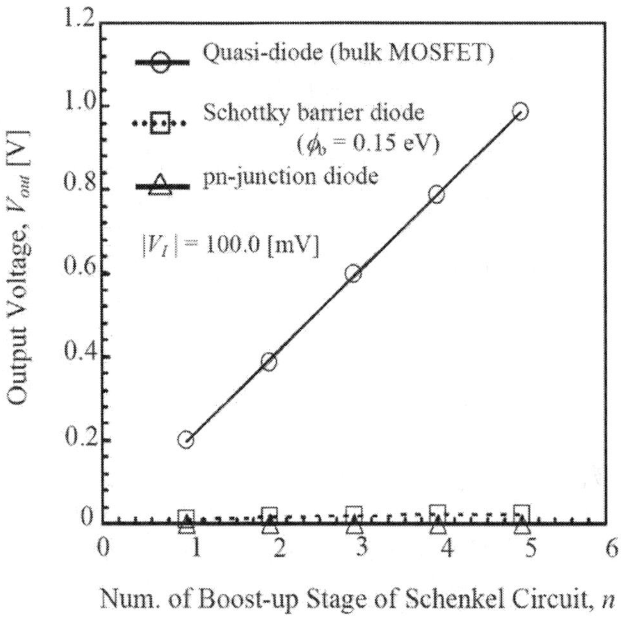

Figure 3. Boost-up performance of Schenkel circuit with various diodes (Pspice simulation results).

In mechanism of voltage boost-up, a large I_R value loses a part of charge stored in a capacitor connected in series. So, the above definition of boost-up efficiency is available independently of the number of boost stages on low-frequency conditions where non-local effects are negligibly small. In addition, the voltage across the capacitor rises when the capacitance of the rectifier is larger that of the capacitor connected in series; however, this results in a low-level current source in contrast to the purpose. Therefore, we have to optimize the capacitance of the capacitor so as to fit the performance request.

Unfortunately, we can not apply the above CB-QD to a practical Schenkel circuit as is because it has a crucial drawback; a CB-QD made on an n-channel bulk MOSFET has a parasitic pn-junction diode between the drain and the substrate that can work when the drain is negatively biased. This effective reverse current I_R of CB-QD that passes through the parasitic pn-junction diode degrades the η value.

Our solution is to base the quasi-diode on an SOI MOSFET instead of a bulk MOSFET to raise the η value. **Figure 4** shows the device structure assumed here and the terminal nodes of an SOI-MOSFET-based quasidiode (SOI-QD). The n-channel fully-depleted (FD)-SOI MOSFET's used here for evaluation of device performance had channel lengths (L) of 0.32 and 1.0 μm (see **Table 1" target="_self"> Table 1**); these devices are used only for a feasibility test and they are not well tempered for the present purpose. Later we perform a. c. simulations for SOI-QD. In this case, it is anticipated that the gate-to-source capacitance and the gate-to-drain capacitance play important roles in the a. c. analyses because they yield various parasitic capacitances including fringing capacitances [14]. So, we consider a realistic device structure to get reliable a. c. simulation results. All physical parameters to draw the cross-section of device are determined on the basis of 0.4-μm CMOS design rule [15]; they are identical to those of the device used in the present experiments. In simulations described later, the doping level of the SOI layer, $N_{a,ch}$, was changed from 6.0×10^{16} cm^{-3} to 3.0×10^{17} cm^{-3} to adjust the threshold voltage.

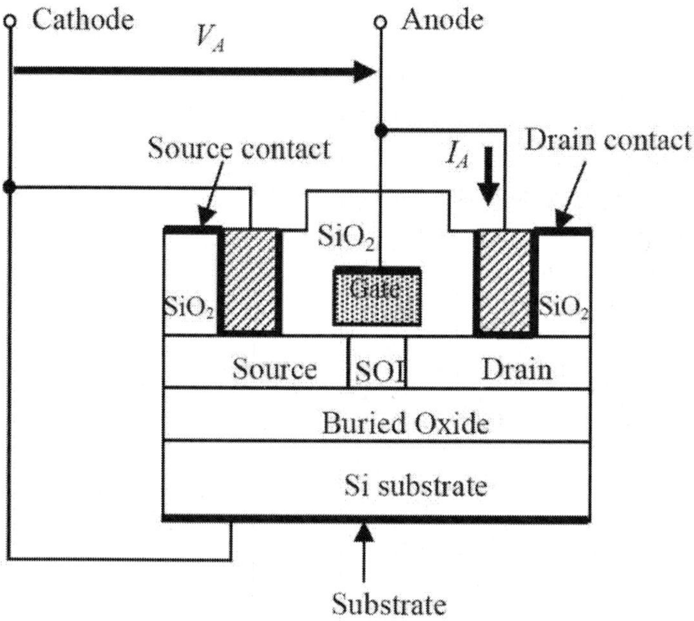

Figure 4. SOI-MOSFET and terminal nodes for quasi-diode operation. Device structure is also assumed for a. c. analyses. Device parameters are carefully designed to take account of practical evaluate gate overlap capacitance, S/D parasitic resistance, fringing capacitance, and other parasitic effects in the device.

Table 1. Device parameters used in experiments.

Parameters	Values [Units]
L	0.32 (or 1.0) [μm]
W	20.6 [μm]
t_{SOI}	50.0 [nm]
t_{BOX}	80.0 [nm]
t_{OX}	7.0 [nm]
$N_{a,ch}$	3.0×10^{17} [cm^{-3}]

Figure 5 shows the I_D-V_G characteristics of the FDSOI MOSFET's (L = 0.32 and 1.0 μm) measured at V_D = 50 mV; the substrate bias was 0 V. The subthreshold swing (S) values of the two devices are quite different. It can be seen that, because of short channel effects, the S value is larger at L = 0.32 μm than at L = 1.0 μm.

Since it is anticipated that the device characteristics are sensitive to substrate bias (V_{SUB}) because of the thin buried oxide layer, we can produce SOI-QD's with various rectification characteristics by modifying the substrate bias (V_{SUB}) applied to the FD-SOI MOSFET. It should be noted that this technique is introduced to examine how the threshold voltage of FD-SOI MOSFET modulates the η value through the change of I_F and I_R. By the substrate bias, we can easily vary the S value and the current level as well as the threshold voltage. As a result, we can find the best solution of the SOI-QD's rectification characteristics to be tuned. In practical applications, we cannot assume the substrate bias to the device because the assumed RF-ID chip has not a voltage supplier; at the stage of device design, the threshold voltage must be tuned by positive voltage parameter if possible. In this paper, as an attempt, we apply a negative, zero or a positive V_{SUB} value to the device; these conditions are labeled by "A", "B" or "C" (see **Table 2" target="_self"> Table 2**), respectively. It should be noted that V_{SUB} value at different L value is not identical when the threshold voltage (V_{TH}) is adjusted to be the same. **Figure 6** shows the I_A-V_A characteristics of SOI-QD "A", "B" and "C".

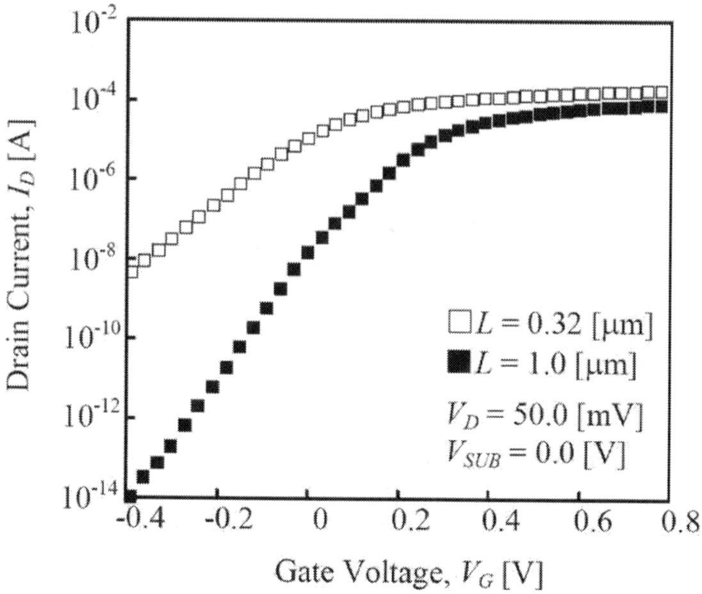

Figure 5. I_D-V_G characteristics of fully-depleted SOI MOSFET (experimental results).

Table 2. I_F, I_R, R_{ch}, and η values of SOI-QD at various operation conditions (*Amplitude of input signal is 100 mV).

	L = 0.32 [μm]	S [mV/dec.]/V_{th} [V]	Forward bias		Reverse bias		η [%]*
			I_F [μA]	R_{ch} [kΩ]	I_R [μA]	R_{ch} [kΩ]	
A	V_{BS} = 0.25[V]	211/0.01	90.0	1.1	−47.0	2.2	64.4
B	V_{BS} = 0.0 [V]	142/0.1	2.1	47.0	−0.89	110	70.7
C	V_{BS} = −1.5 [V]	132/0.27	1.2	81.0	−0.55	180	69.3
	L = 1.0 [μm]	S [mV/dec.]/V_{th} [V]	Forward bias		Reverse bias		η [%]*
			I_F [μA]	R_{ch} [kΩ]	I_R [μA]	R_{ch} [MΩ]	
A	V_{BS} = 3.0 [V]	169/0.01	29.0	3.4	−11.0	8.8e−3	72.0
B	V_{BS} = 0.0 [V]	71.8/0.20	0.25	401.0	−0.016	6.2	94.0
C	V_{BS} = −3.0 [V]	71.5/0.92	7.2e−3	1.39e4	−2.6e−4	3.9e3	96.6

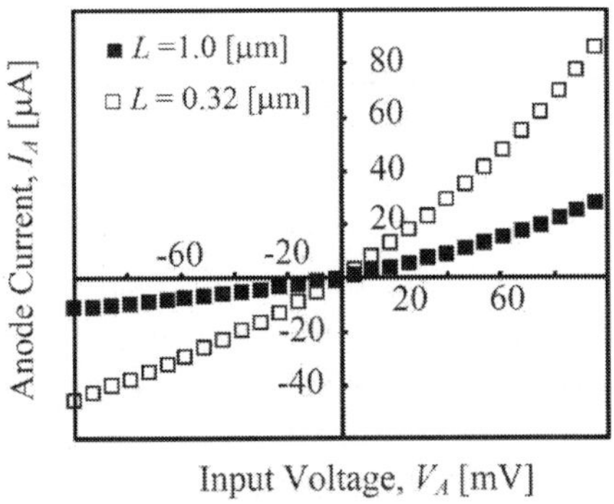

(a)

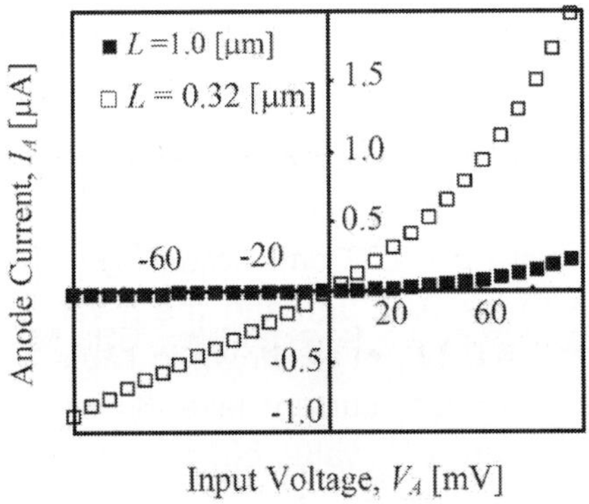

(b)

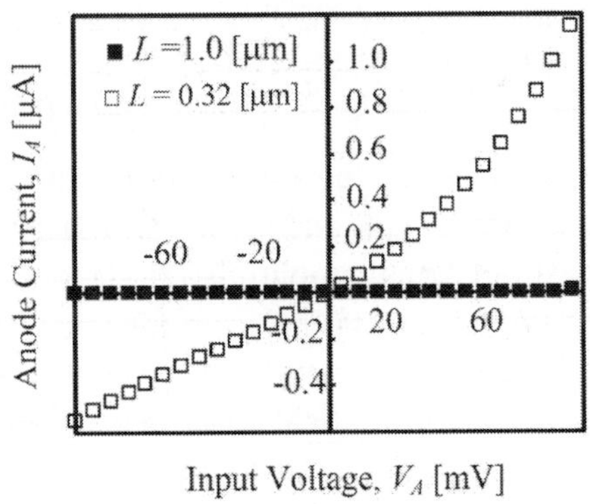

(c)

Figure 6. Rectifier I-V characteristics of SOI-QD (experimental results). (a) Operation condition "A" ($V_{SUB} > 0$); (b) Operation condition "B" ($V_{SUB} = 0$ V); (c) Operation condition "C" ($V_{SUB} < 0$).

At first, we consider the impact of L value as shown in **Figure 6**. In condition "A", the SOI-QD works near the threshold voltage (V_{TH}) because V_{SUB} is positive. S value is large in condition "A" than in condition "B", resulting in a smaller ratio of I_F/I_R and thus a smaller η value (see **Table 2**, and**Figures 6**(a) and (b)). In condition "A", however, since the channel resistance (R_{ch}) is reduced due to the lowering of the threshold voltage (V_{TH}), the driving current of the device increases. In contrast, in condition "C", the device works in the subthreshold region because of the negative substrate bias (V_{SUB}). The S value is smaller in condition "C" than in condition "B"; the channel resistance (R_{ch}) in condition "C" increases due to the raising of the threshold voltage (V_{TH}), resulting in a lower drive current, but identical η value to that seen in condition "B" (see **Figure 6**(c)).

Next we compare the performance of the two devices (L = 1.0 and 0.32 μm) shown in **Figure 6**. As mentioned above, the S value with L = 1.0 μm is smaller than that with L = 0.32 μm. Accordingly, I_F/I_R is larger with L = 1.0 μm than that with L = 0.32 μm, and the η value with L = 1.0 μm is larger than that with L = 0.32 μm as shown in **Table 2**. It should be noted that the difference in forward current level (I_F) at the same bias comes from not only the different W/L value but also the different S value. This means that we must take account of the trade-off between η and I_F; that is, it is necessary to select the most suitable operation bias condition and to reduce the S value. **Table 2**suggests that the S value should be less than 75 mV/dec to reduce I_R value and threshold voltage should be ranging from 0.1 V to 0.2 V to simultaneously gain a high η value.

Finally, we briefly compare the performance of SBD, pn-junction diode and SOI-QD; **Table 2** shows I_F, I_R and η of SOI-QD at various operation conditions when the input signal amplitude ($|V_A|$) is 100 mV. According to a recent report [3], when V_A = 100 mV, the SBD has an I_F value of about 1.0 nA and the pn-junction diode has an I_F value of about 0.1 pA; this indicates that the SOI-QD has identical I_F to the other devices or has larger I_F at all conditions for the two L values examined. In addition, since the SOI MOSFET has lower leakage current than the SBD and subthreshold characteristic of well-tempered SOI MOSFET is excellent, a high boost efficiency can be expected when the SOI-QD is used.

SIMULATION-BASED CONSIDERATION OF RF PERFORMANCE OF SOI-QD

In order to investigate of the feasibility of using SOIQD's in RF applications [16], we conducted extensive a. c. analyses using a 2D device simulator [11]. **Figure 7" target="_self"> Figure 7**shows the simulated rectifier characteristics of the SOIQD for various $N_{a,ch}$ values. Since the threshold voltage (V_{th}) rises sharply with the increase in doping level of the SOI layer ($N_{a,ch}$), the forward-biased anode current (I_F) decreases greatly and the reverse-biased anode current (I_R) also decreases. However, η increases on the basis of Equation (1) as $N_{a,ch}$ increases because the reduction of I_R overwhelms that of I_F, which is as is expected. A simple estimation method of η value of SOI MOSFET is shown in Appendix B.

When a high-frequency operation is considered, a. c. response to applied signal amplitude should be evaluated; this is very important in SOI MOSFET because it is anticipated that the floating-body effect delays the current response to the applied signal [17]. We think that anode conductance $g_A(=dI_A/dV_A)$ successfully traces response capability of SOI MOSFET because g_A is extracted from a small signal analysis; we think a large signal analysis is not always required because Fourier transformation results of signals are effectively considered. Then we define the boost-up efficiency to a. c. signals as

$$\eta = \frac{\{\Delta I_{AF}(f)/\Delta V_A(f)\}}{\{\Delta I_{AF}(f)/\Delta V_A(f) + \Delta I_{AR}(f)/\Delta V_A(f)\}} = g_{AF}(f)/\{g_{AF}(f) + |g_{AR}(f)|\} \quad , \qquad (2)$$

where g_{AF} and g_{AR} mean the anode conductance at the forward and the backward bias conditions, respectively;

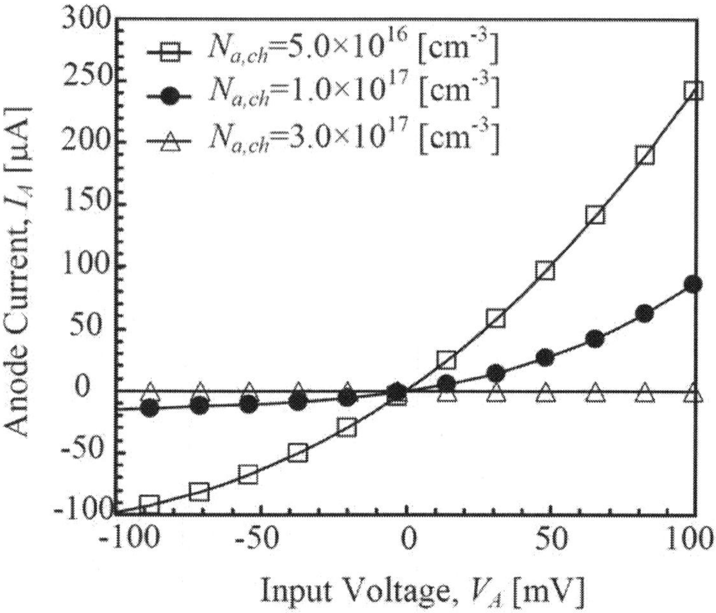

Figure 7. Simulated I-V characteristics of quasi-diodes (device simulations).

I_{AF} and I_{AR} are the anode current at the forward and the backward bias conditions, respectively. Some **Figure 8** shows the simulated g_A-V_A curves for various $N_{a,ch}$ values, where g_A (=dI_A/dV_A).**Figure 8**(a) is for $N_{a,ch}$ = 5 × 10^{16} cm^{-3}, **Figure 8**(b) for $N_{a,ch}$ = 1 × 10^{17} cm^{-3}, and **Figure 8**(c) for $N_{a,ch}$ = 3 × 10^{17} cm^{-3}. Frequency f was changed from 1 Hz to 10 GHz. In the case of $N_{a,ch}$ = 5.0 × 10^{16} cm^{-3} (see **Figure 8**(a)), the g_A-V_A characteristic is not sensitive to frequency (1 Hz to 10 GHz). In the cases of $N_{a,ch}$ = 1.0 × 10^{17} cm^{-3} (see **Figure 8**(b)) and 3.0 × 10^{17} cm^{-3} (see**Figure 8**(c)), however, the g_A-V_A characteristic is sensitive to frequency. In particular, for $N_{a,ch}$ = 3.0 × 10^{17} cm^{-3}, the g_A-V_A characteristic reacts strongly to frequency.

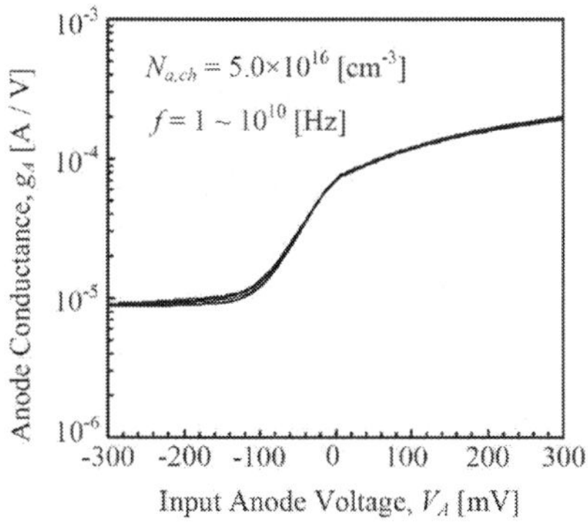

(a)

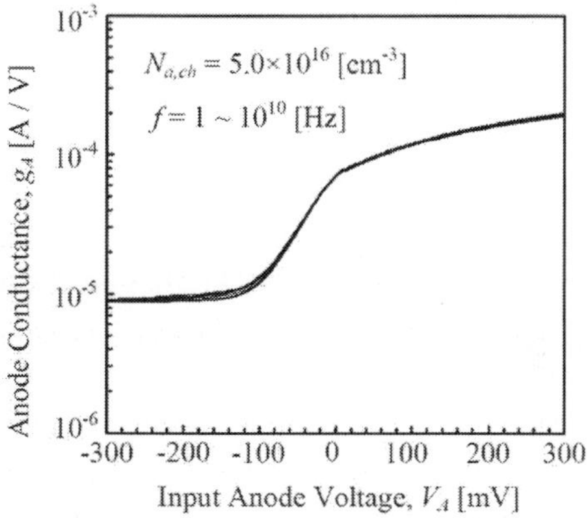

(b)

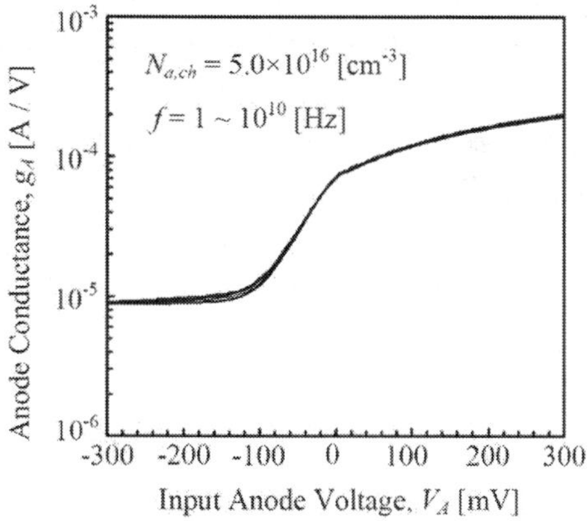

(c)

Figure 8. Simulated g_A-V_A characteristics (device simulations). (a) $N_{a,ch} = 5.0 \times 10^{16}$ [cm^{-3}]; (b) $N_{a,ch} = 1.0 \times 10^{17}$ [cm^{-3}]; (c) $N_{a,ch} = 3.0 \times 10^{17}$ [cm^{-3}].

Since the SOI layer is 50 nm thick, the fully-depleted condition is satisfied in two cases (**Figures 8**(a) and (b)) [18]. Majority carriers (holes) are basically not responsible for device operation, and so the parasitic bipolar action is not expected at the bias condition used ($|V_A| < 0.3$ V). Since the threshold voltage is very low (~0 V), electrons in the inversion layer rule device operation; in this case, the frequency dependence of the dielectric response of electrons ("majority carriers" near the surface) limits the g_A-V_A characteristic. The limit of the frequency response of electrons in Si is higher than 100 GHz [16], so the simulation results shown in **Figures 8**(a) and (b) are acceptable. The smaller variation in g_A-V_A characteristics seen in **Figure 8**(b) is related to the remaining hole density near the bottom of SOI layer, which should be higher than that in **Figure 8**(a).

We note that the fully-depleted condition is not satisfied in the case of $N_{a,ch} = 3.0 \times 10^{17}$ cm^{-3}[18]. That is, the majority carriers (holes), remaining near the SOI/ buried oxide interface, play an important role in determining device operation, which corresponds to the typical floating body effect [17]. Since the g_A-V_A characteristics at frequencies above 10 MHz differ from those below 10 MHz as seen in **Figure 8**(c); in other words, the g_A-V_A characteristics at frequencies above 10 MHz are not normal. In the present case, it is easily anticipated that the generation process of majority carriers (holes) (around the junction and inside the depleted body) rules the dynamic operation of SOI-QD. Since the generationrecombination time constant is about 0.1 μsec in the present simulations (see Appendix A), the hole generation process does not respond at frequencies above 10 MHz, resulting in the body floating effect [17].

Figure 9 shows the simulated $\eta - f$ characteristics of the SOI-QD from f = 1 Hz to 1 THz with the parameter of $N_{a,ch}$, although the simulated value of η is not reliable for $f > 100$ GHz because physical models for devices in the device simulator [11] are not proposed for such a high frequency; we simply focus on the behavior of η. η values shown in **Figure 9** are calculated by Equation (2).

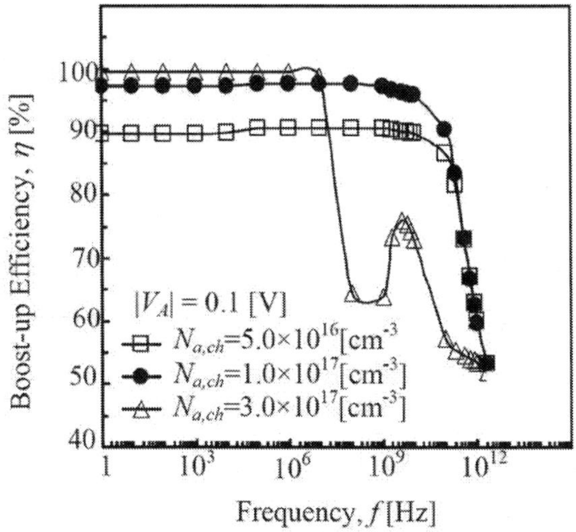

Figure 9. η-f characteristics for various body doping levels (device simulations).

In **Figure 9**, it should be noted that η values calculated by g_A (Equation (2)) at low frequency region are almost identical to η values calculated from d. c. current (experimental results) for L = 1.0 μm by Equation (1). So this suggests that η value can be estimated using Equation (2) at RF region. In **Figure 9**, we can see that the η value remains higher than 90% independently of $N_{a,ch}$up to 10 MHz; this is supported by the fact that S value in the active range of I_A of SOI-QD is sufficiently small. However, when f is higher than 100 MHz, especially when $N_{a,ch}$ = 3.0 × 10^{17} cm^{-3}, η falls to 60%. This comes from the body-floating effect as mentioned previously. On the other hand, at $N_{a,ch}$ = 5.0 × 10^{16} cm^{-3} and 1.0 × 10^{17} cm^{-3}, η remains high up to 10 GHz. As a result, SOI-QD's with $N_{a,ch}$ = 5.0 × 10^{16} cm^{-3} and 1.0 × 10^{17} cm^{-3} can be used in RF applications; when $N_{a,ch}$ = 3.0 × 10^{17} cm^{-3}, the SOI-QD is no longer suitable because of the significant body floating effect.

Figure 10 shows the simulated η-$N_{a,ch}$ characteristics of the SOI-QD at 1 MHz and 4 GHz. In the low frequency range (~1 MHz) η slightly increases with $N_{a,ch}$ because of the reduction of S at the driving point, and finally reaches its upper limit when the reduction in S value ceases. Since $N_{a,ch}$increases, driving current (I_A) decreases because the threshold voltage of FD-SOI MOSFET rises. Thus, the doping level of the SOI layer of SOI-QD must be optimized to realize a high performance Schenkel circuit. In the radio frequency range (~4 GHz), η peaks because of the floating body effect mentioned above. This strongly suggests that the value of $N_{a,ch}$ must be selected so as to hold the SOI layer in the fully-depleted condition of the SOI layer. As for the present simulation, $N_{a,ch} = 1.0 \times 10^{17}$ cm^{-3} yields the best SOI-QD performance in RF applications up to 100 GHz; when SOI layer is thinned to a range of sub-50 nm, $N_{a,ch}$ value higher than 1×10^{17} cm^{-3} can be applied to devices [19].

In the above simulations, we assumed $V_A = 0.1$ V because we considered the case of short-distance communications. Since the above simulations show that the pro-

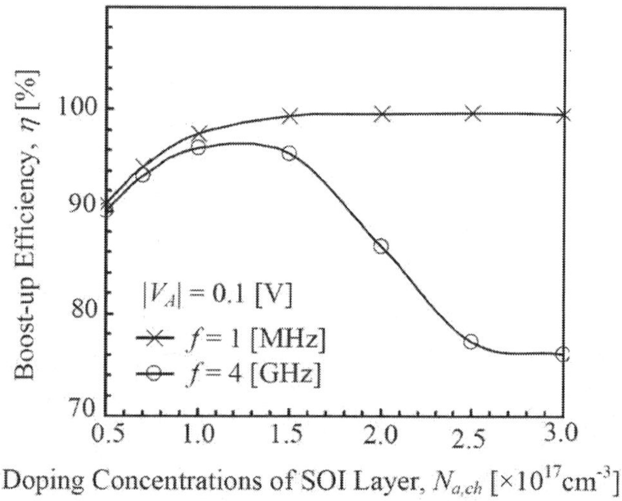

Figure 10. η-$N_{a,ch}$ characteristics for various frequencies (device simulations).

posed SOI-QD produces almost identical performance at $V_A < 0.1$ V, we think that Schenkel circuits with SOI-QD's are also applicable to long-distance communications.

CONCLUSION

The feasibility replacing SBD's in the Schenkel circuit with SOI-MOSFET's as quasi-diodes was examined by experiments and simulations. The reverse-biased current (I_R) of the SOI-QD is much lower than its forward-biased current (I_F) and the driving current (I_F) is high because of the excellent S value provided by the SOI-MOSFET arrangement; we noted that the trade-off between boost efficiency (η) and I_F should be taken into account. In addition, a. c. analyses using a two-dimensional device simulator showed that the body doping concentration ($N_{a,ch}$) of the SOI layer should be optimized so as to hold the fully-depleted condition for RF applications up to 100 GHz.

ACKNOWLEDGEMENTS

This study is financially supported by Kansai University research grants, Grant-in-Aid for Joint Research (2004-2005). The authors wish to express their thanks to Mr. Takuta Tamura, Mr. Yuki Tahara, and Mr. S. Namura for their technical support on device simulations.

REFERENCES

1. U. Karthaus and M. Fischer, "Fully Integrated Passive UHF RFID Transponder IC with 16.7-μW Minimum RF Input Power," IEEE Journal of Solid-State Circuits, Vol. 38, No. 10, 2003, pp.1602-1608.
2. M. Usami and M. Ohki, "The μ-Chip: Ultra-Small 2.45 GHz RFID Chip for Ubiquitous Recognition Applications," IEICE Transactions on Electronics, Vol. E86-C, No. 4, 2003, pp.521-528.
3. W. Jeon, T. M. Firestone, J. C. Rodgers and J. Melngailis, "Design and Fabrication of Schottky Diode, On-Chip RF Power Detector," Solid-State Electronics, Vol. 48, No. 10-11, 2004, pp. 2089-2093.
4. B. Strassner and K. Chang, "Passive 5.8-GHz Radio-Frequency Identification Tag for Monitoring Oil Drill Pipe," IEEE Transactions on Microwave Theory and Techniques, Vol. 51, No. 2, 2003, pp. 356-363. doi:10.1109/TMTT.2002.807832
5. K. Ahsan, H. Shah and P. Kingston, "RFID Applications: An Introductory and Exploratory Study," International Journal of Computer Science Issues, Vol. 7, No. 1, 2010, pp. 1-7.
6. Y. Kado, M. Suzuki, K. Koike, Y. Omura and K. Izumi, "A 1 GHz/0.9 mW CMOS/SIMOX Divide-by-128/129 Dual-Modulus Prescaler Using a Divide-by-2/3 Synchronous Counter," IEEE Journal of Solid-State Circuits, Vol. 28, No. 4, 1993, pp. 513-517.
7. O. Rozeau, J. Jomaah, J. Boussey and Y. Omura, "Comparison between Highand Low-Dose Separation by Implanted Oxygen MOS Transistors for Low-Power RadioFrequency Applications," Japanese Journal of Applied Physics, Vol. 39, No. 4B, 2000, pp. 2264-2267.doi:10.1143/JJAP.39.2264
8. J. P. Raskin, A. Viviani, D. Flandre and J.-P. Colinge, "Substrate Crosstalk Reduction Using SOI Technology," IEEE Transactions on Electron Devices, Vol. 44, No. 12, 1997, pp. 2252-2261. doi:10.1109/16.644646
9. Y. Omura, "Negative Conductance Properties in Extremely Thin Silicon-on-Insulator Insulated-Gate pn-Junction Devices (Silicon-on-Insulator Surface

Tunnel Transistor)," Japanese Journal of Applied Physics, Vol. 35, No. 11A, 1996, pp. L1401-L1403.doi:10.1143/JJAP.35.L1401

10. Y. Omura and T. Tochio, "Significant Aspects of Minority Carrier Injection in Dynamic-Threshold SOI MOSFET at Low Temperature," Cryogenics, Vol. 49, No. 11, 2009, pp. 611-614. doi:10.1016/j.cryogenics.2008.11.009
11. Synopsys Inc., "TCAD-DESSIS/GENESISe Operation Manual," ver. 7.5. http://www.sysnopsys.com/
12. Microsim Corp., "PSpice Reference Manual." http://www.microsimcom.com/
13. K. Takahashi, S. Y. Wang and M. Mizunuma, "Complementary Charge Pump Booster," Electronics and Communications in Japan Part II-Electronics, Vol. 82, No. 6, 1999, pp. 73-81. doi:10.1002/(SICI)1520-6432(199906)82:6<73::AID-ECJB8>3.0.CO;2-C
14. Y. Taur and T. H. Ning, "Fundamentals of Modern VLSI Devices," Cambridge University Press, Cambridge, 1998.
15. Y. Kado, M. Suzuki, K. Koike, Y. Omura and K. Izumi, "An Experimental Full-CMOS Multi-gigahertz PLL LSI Using 0.4-µm Gate Ultrathin-Film SIMOX Technology," IEICE Transactions on Electronics, Vol. E76-C, No. 4, 1993, pp. 562-571.
16. C. Wann, F. Assaderaghi, L. Shi, K. Chan, S. Cohen, H. Hovel, K. Jenkins, Y. Lee, D. Sadana, R. Viswanathan, S. Wind and Y. Taur, "High-Performance 0.07-µm CMOS with 9.3-ps Gate Delay and 150 GHz f_T," IEEE Electron Devices Letters, Vol. 18, No. 12, 1997, pp. 625-627. doi:10.1109/55.644091
17. J.-P. Colinge, "Silicon-on-Insulator: Materials to VLSI," 3rd Edition, Kluwer Academic Publishing, Dordrecht, 2004. doi:10.1007/978-1-4419-9106-5
18. Y. Omura, S. Nakashima and K. Izumi, "Investigation on High-Speed Performance of 0.1-µm-Gate, Ultrathin-Film CMOS/SIMOX," IEICE Transactions on Electronics, Vol. E75-C, No. 12, 1992, pp. 1491-1497.
19. Y. Omura, S. Nakashima, K. Izumi and T. Ishii, "0.1-µmGate, Ultrathin-Film CMOS Devices Using SIMOX Substrate with 80-nm-Thick Buried Oxide Layer," IEEE Transactions on Electron Devices, Vol. 40, No. 5, 1993, pp. 1019-1022. doi:10.1109/16.210214
20. J. R. Brews, "A Charge-Sheet Model for the MOSFET," Solid State Electronics, Vol. 21, No. 2, 1978, pp. 345-355. doi:10.1016/0038-1101(78)90264-2
21. Y. Sakurai, A. Matsuzawa and T. Dozeki, Eds., "FullyDepleted SOI CMOS Circuits and Technology for Ultralow-Power Applications," Springer, Berlin, 2006.

APPENDIX A: SIMULATION MODEL FOR MINORITY CARRIER LIFETIME

Here we introduce the model for minority carrier lifetimes [11] used in PSPICE simulations to ensure consistency with DESSIS simulation results.

$$\tau = \tau_{\max} + \frac{\tau_{\max} - \tau_{\min}}{1 + \left(N / N_{ref} \right)^{\gamma}}, \quad \text{(A1)}$$

where N is the doping density, N_{ref} is the doping parameter, γ is the fitting parameter, τ_{max} and τ_{min} are lifetime parameters. Parameter values used here are summarized in **Table 3**.

APPENDIX B: ON THE DESIGN GUIDELINE OF SOI-QD'S

We rewrite the expression of boost-up efficiency given by Equation (1).

$$\eta = I_F / \left(I_F + |I_R| \right), \quad \text{(B1)}$$

When no short-channel effect is assumed, I_F and I_R can be expressed approximately as

$$I_F = \left(\frac{W}{2(1+\alpha)L} \right) \mu_n C_{ox} (V_A - V_{TH})^2 + I_{TH} \quad (V_A > V_{TH}), \quad \text{(B2)}$$

$$I_R = I_{TH} 10^{\frac{V_A - V_{TH}}{S}} \quad (V_A < V_{TH}), \quad \text{(B3)}$$

where most of notations are conventional, and S stands for the subthreshold swing for the fully-depleted SOI MOSFET [17]. Parameters α and I_{TH} (threshold current) are given by

$$\alpha = \frac{C_{\mathrm{SOI}} C_{\mathrm{BOX}}}{C_{ox} \left(C_{\mathrm{SOI}} + C_{\mathrm{BOX}} \right)}, \quad \text{(B4)}$$

$$I_{TH} = \mu_n \left(\frac{W}{L} \right) qN_A \left(\frac{kT}{q} \right)^2 \frac{1}{E_S}, \tag{B5}$$

where C_{SOI} is the SOI layer capacitance [17], C_{BOX} is the buried oxide layer capacitance [17], and E_s is the surface electric field. The derivation of α is given in [17], and that of I_{TH} is given in [20,21].

Table 3. Physical parameters assumed in device simulations (DESSIS).

Parameters	Values [units]	Comments
τ_{max}	1.0×10^{-5} [s]/3.0×10^{-5} [s]	Electrons/Holes
τ_{min}	0.0 [s]/0.0 [s]	Electrons/Holes
N_{ref}	1.0×10^{16} [cm^{-3}]	-
γ	1.0	-

CITATION

Y. Omura and Y. Iida, "Performance Prospects of Fully-Depleted SOI MOSFET-Based Diodes Applied to Schenkel Circuit for RF-ID Chips," Circuits and Systems, Vol. 4 No. 2, 2013, pp. 173-180. doi: 10.4236/cs.2013.42024.

Chapter 1

A Parallel Circuit Simulator for Iterative Power Grids Optimization System

Taiki Hashizume[1], *Masaya Yoshikawa*[2], *Masahiro Fukui*[1]

[1]Department of VLSI System Design, Ritsumeikan University, Kusatsu, Japan
[2]Department of Information Engineering, Meijo University, Nagoya, Japan

ABSTRACT

This paper discusses a high efficient parallel circuit simulator for iterative power grid optimization. The simulator is implemented by FPGA. We focus particularly on the following points: 1) Selection of the analysis method for power grid optimization, the proposed simulator introduces hardware-oriented fixed point arithmetic instead of floating point arithmetic. It accomplishes the high accuracy by selecting appropriate time step of the simulation; 2) The simulator achieves high speed simulation by developing dedicated hardware and adopting parallel processing. Experiments prove that the proposed simulator using 80 MHz FPGA and eight parallel processing achieves 35 times faster simulation than software processing with 2.8 GHz CPU while maintaining almost same accuracy in comparison with SPICE simulation.

INTRODUCTION

With the deep submicron technologies, it has become possible to mount a large and high performance system on one VLSI chip. However, the power supply voltage is lowering along with shrinking of the device size. On the other hand, the power consumption is increasing because the number of transistors and the clock frequency increase. Therefore, IR-drop and electro-migration (EM) in the power grids make functions and devices unreliable. It becomes more serious problem than ever. The power grid optimization has become very important for ensuring the reliability, correctness and stability of the design. In recent years, many prior researches [1-6] have been proposed to solve the problem. A heavy simulation time is required to analyze the problem of heat and electromagnetic field [3]. Moreover, a high speed and high accurate circuits simulator is required for iterative optimization which includes execution of the simulation many times for its evaluation.

Thus, several hardware accelerator approaches have been reported [7-12]. Lee et al. achieved 25 times high speed processing in the timing verification using multi CPU [7]. Nakasato et al. demonstrated particle simulation by floating point arithmetic on an FPGA [8]. Watanabe et al. introduced parallel-distributed time-domain circuit simulation of power distribution networks by multiple PC [9]. However, no research of hardware accelerator to optimize power grids of VLSI has been reported.

This paper aims at estimating the performance improvement by hardware implementation of the circuit simulation in the power supply wiring. We have analyzed accuracy, speed, and hardware resources to implement typical numerical analysis algorithms of circuit simulation, in cases of floating point and fixed point variables, software and hardware. Moreover, we propose an efficient hardware circuit simulator for power grid optimization. It achieves high speed simulation by adopting pipeline and parallel processing. The proposed simulator introduces hardware-oriented fixed point arithmetic instead of floating point arithmetic. The hardware-oriented fixed point arithmetic realizes the high area-efficiency by reducing hardware resources, and it also accomplishes the high accuracy by controlling intervals of simulation.

POWER GRID OPTIMIZATION SYSTEM

Power Grid Optimization Algorithm

The power grid consists of two vertical and horizontal layers. Those are interconnected by contacts at the intersection. The equivalent circuit model for proposed power grid optimization system is shown in **Figure 1**.

Decoupling capacitors are inserted to decrease dynamic IR-drop and inductor noise to the power grid. At each grid area, four edges of resistances are placed. The capacitance which connected to each node represents the sum of the decoupling capacitance and the wiring capacitance.

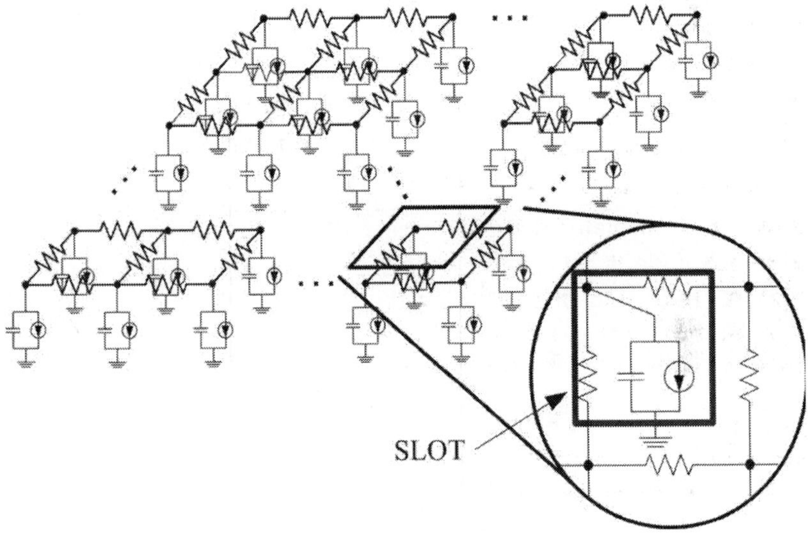

Figure 1. Equivalent circuit model for power grids.

Current source which connected to each node corresponds to the current consumption of functional block of VLSI. These circuit elements connected with a grid are defined as SLOT. The power grid optimization is defined as a multi-objective optimization problem. First objective is to reduce the risk of timing error which is caused by transient and local IR-drop. Second objective is to reduce the risk of wire break which is caused by excess of electric current density at a portion of a wiring. Third objective is to reduce the risk of failing signal wiring which is caused by local congestion of wires. We have already proposed an algorithm to solve these trade-off problems [13-15]. It defines the optimization problem by using an evaluation function which unifies the risks of multiple objectives. The optimization is scheduled by an iterative improvement. Iterative operation consists of circuit simulation, evaluation of risks, and small modification of wire width or decupling capacitance.

Organization of Power Grid Optimization System

The power grid optimization system consists of three parts, the simulation part, the evaluation part and the optimization part, as shown in **Figure 2**. A new repetitive optimization and evaluation function is introduced to improve the multiple physical issues [13]. To evaluate circuits, an original metrics, RISK, is defined. The proposed power grid optimization system adopts the algorithm as a base algorithm of the optimization part. Effectiveness of the power grid optimization system is determined by communication-overhead between each part, in addition to the simulation accuracy and the processing speed.

The simulation part is the most time consuming. It simulates the circuit behavior and calculates the electric current of each edge and the voltage of the each node. It takes more than 99% of the total computation time. Therefore, reduction of processing time of the simulation part is most important from the point of high speed optimization processing. The evaluation part defines a risk function for each design metric, i.e., IR-drop, EM, and wiring congestion, from simulation results.

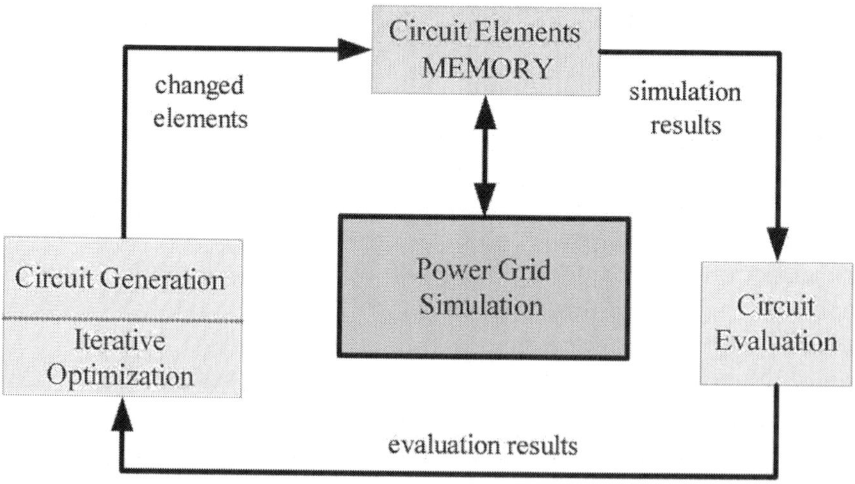

Figure 2. Power grid optimization system.

IR-drop RISK is defined as a probability of causing the timing error. An increase of current density raises the EM RISK and it cause a deterioration and disconnection of wires to be unable to operate. As a result, the circuit doesn't operate. The wiring congestion RISK represents the proportion of the wiring resource in a SLOT area. It is a restriction of preventing from an excessive wiring and state that cannot be wired. Because each grid is composed of the power grid, decoupling capacitor and signal wire, totals of those areas are compared with the grid area.

The optimization part improves IR-drop, EM and wiring congestion by changing wiring width and decoupling capacitance. Performance of the power grid optimization system is determined by communication overhead between each part, in addition to the simulation accuracy and the processing speed.

Simulation Algorithm

Circuit Analysis Method

This section summarizes typical numerical analysis methods which can be used for the power grid simulation. Euler method and Runge-Kutta method are typical numeric method to solve differential equations. Generally, small time step must be selected to analyze steadily by Forward Euler method (FE).

The computation algorithm is simple and needs smaller hardware resources, but it may easily diverge. RungeKutta methods require complex calculation, but the simulation is more stable even though we select a larger time step than FE. Thus, we must carefully select the numerical algorithm for the given problem. We have examined 100 or more variations of power grids and evaluated.

The test data is composed of four size variations, 50 × 50, 100 × 100, 200 × 200, 500 × 500. Also, it is composed of five RC distributions, regular, random, two types of hand-specified, extracted from real chips. The largest RC time constant was about 20 times larger than the minimum one.

As the results, the accuracy is almost same for each numerical method and we have selected FE. For the evaluation of numerical methods see the next section.

Evaluation of the Accuracy by Comparing with SPICE

This In order to verify the proposed algorithm's validity, comparative experiments are performed using the following two types of circuit simulations: 1) Accuracy comparison between double-precision floating point arithmetic and SPICE; and 2) Accuracy comparison between single-precision fixed point arithmetic and double-precision floating point arithmetic. The power grid which is used for the analysis is a structure of uniform RC distribution, and R and C in the power grid were randomly set from predefined three values. The power grid scale is 10 × 10 grids. The dynamic current consumption is changed in every 10 [μsec]. In conversion into the fixed point, the fraction part of all variables was set to 22-bit. Since 32-bit adders and multipliers are used for experiment, the overflow happened in 23-bit or more.

Firstly, the result of experiments on each simulation with double-precision arithmetic is shown in Figures 3 and 4. The step size is an important point in the transition analysis, and it is necessary to set it to small step to execute an accurate analysis. The small step is set 0.63 [μsec]. The smallest RC was 20 [μsec]. All the analysis methods have been achieved high accuracy compared with SPICE.

Next, experiments on fixed point arithmetic are executed by various step sizes, and the error is verified with the floating point arithmetic. In small step size, the maximum error margin and the average error margin are shown in Tables 1 and 2.

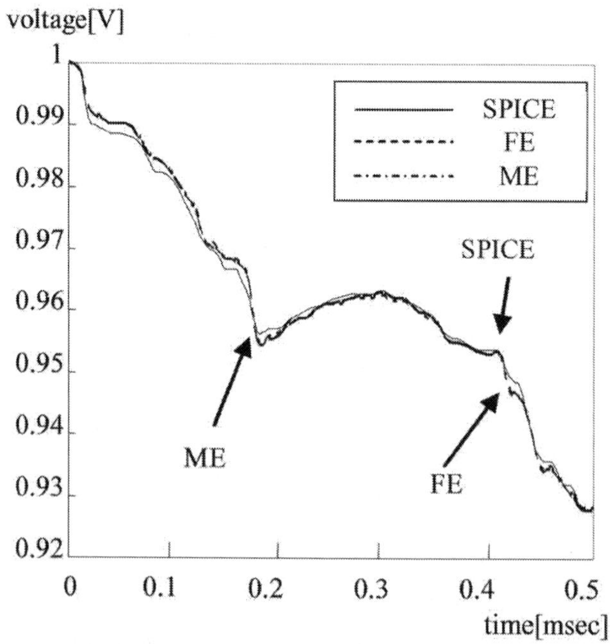

Figure 3. Simulation result by Euler method with small step size.

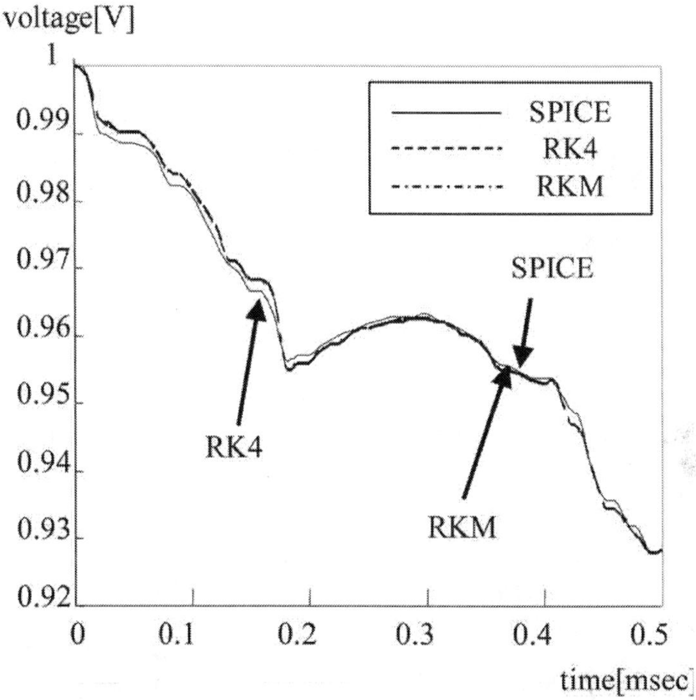

Figure 4. Simulation result by Euler method with small step size.

Table 1. Error by fixed point arithmetic for uniform RC circuit.

	FE	ME	RK4	RKM
max [%]	0.11	0.11	0.11	0.11
Ave [%]	0.021	0.021	0.020	0.020

Table 2. Error by fixed point arithmetic for random RC circuit.

	FE	ME	RK4	RKM
max [%]	0.057	0.057	0.058	0.057
Ave [%]	0.0083	0.0083	0.0083	0.0082

In the uniform RC distribution, all nine patterns, the combination of the wiring resistance and the capacitance, are executed. **Table 2** shows the maximum error and the average error of 100 kinds of circuit selected at random. The high accurate simulation has been achieved with a small time step.

In fixed point arithmetic, fourth order Runge-Kutta Method (RK4) and Runge-Kutta marsun method (RKM) obtain a good result. In addition, the error of uniform RC distribution is larger than that of random circuit. The computational complexity of FE and Modified Euler method (ME) are small. In contrast, RK4 and RKM are complex for computation, though they can execute an accurate analysis.

In our preliminary experiments, FE and RK4 diverged on the same time step. Therefore, FE is superior to RK4 in case of the power grid optimization problem.

Simulation Flow

Voltage and current change in each node at small time step is analyzed based on information of the RC distribution, the current consumption distribution, and the supply voltage, etc. The simulation flow is shown in **Figure 5**, and a part of the equivalent circuit is shown in **Figure 6**.

In the simulation part, the current of each wiring is calculated from voltage distribution and the wiring resistance. Then, the charge which accumulates in the capacitance connected with each node is calculated, and the voltage is derived every small time step dT. These processing are iterated during simulation time for entire power grid T_m, and assumed to be end of simulation. To obtain the voltage at each node, the charge is changed by the inflow current I_{left} and I_{up}, the outflow current I_{right} and I_{down}, and the current consumption I_{con}. Voltage and current change in each node are computed based on RC distribution at small time step.

For more high speed simulation, the simulation with hardware is effective, but the achievement of the hardware simulator is not easy according to the restriction of the error margin and the hardware resource. To examine whether it is feasible to make the present simulation hardware, the accuracy of the analysis by the doubleprecision floating point arithmetic and by the single-precision fixed point arithmetic are verified. The error margin is caused by replacing with fixed point arithmetic. It is because the fixed point arithmetic can be processed at the same speed as the integer operation. Additionally, the area of the fixed point arithmetic unit is far smaller compared with the floating point arithmetic unit.

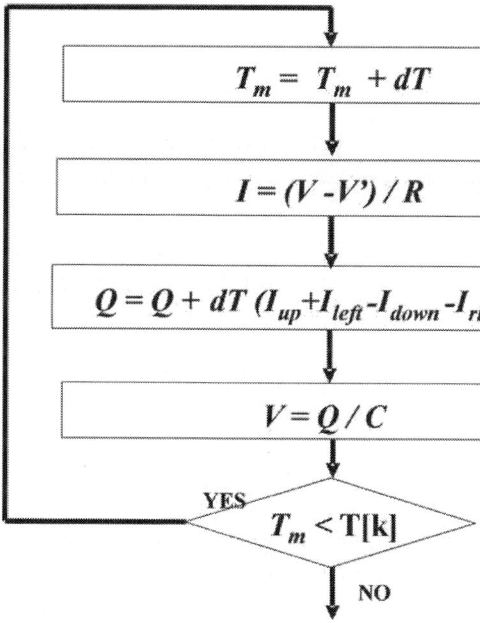

Figure 5. Simulation flow.

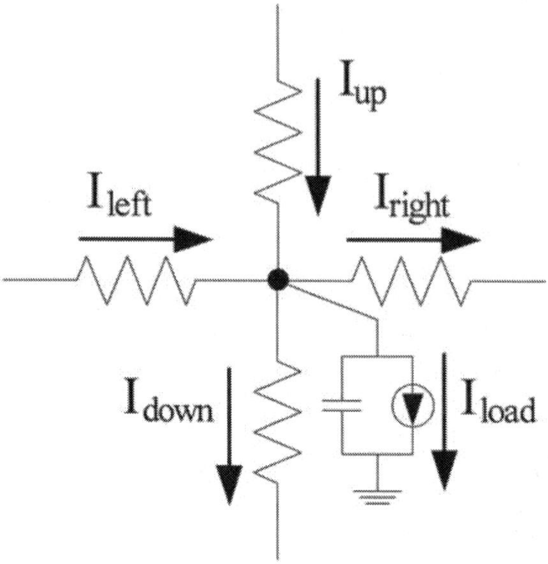

Figure 6. A part of equivalent circuit.

HARDWARE ARCHITECTURE FOR POWER GRID SIMULATION

The proposed simulation algorithm adopts fixed point arithmetic. It achieves the same processing speed as the integer operation, and has an advantage of area-efficiency when implementing on hardware. Fixed point arithmetic includes a risk of overflow; however, this risk is reduced by correction processing and bit shift. The simulation part computes the voltage and the current of each node of power grid, and stores the simulation results into memory. Furthermore, it is necessary to control the simulation and the memory behavior to achieve high speed simulation.**Figure 7** shows the block diagram of the power grid simulation which corresponds to simulation module and memory access as shown in **Figure 2**. Control module controls state transition and timing of each module. Each circuit variables are stored in each memory. Wiring resistance and capacitance are updated when changing the circuit design.

The voltage and the charge are updated whenever advancing at a small time step. Therefore, these are stored in RAM. Circuit variables are fetched from the memory into the simulation module, and the results are written into memory.

The simulation module is composed by adder, subtractor, and multiplier, and doesn't use divider. Current is calculated by dividing potential difference of wiring resistance, however divider needs a lot of implementation areas. Therefore, divider is converted into multiplier by storing each reciprocal when wiring resistance and capacitance are stored in the memory. **Figure 8** shows the calculation procedure in the simulation module.

Input variables, "G_{right}", "G_{down}" and "Z" show the reciprocal of wiring resistance and capacitance respectively. In the current calculation of the hardware algorithm, only outflow current is computed as shown in **Figure 4**. Because inflow current of a point corresponds to outflow current of the adjacent point (left side or upper side).

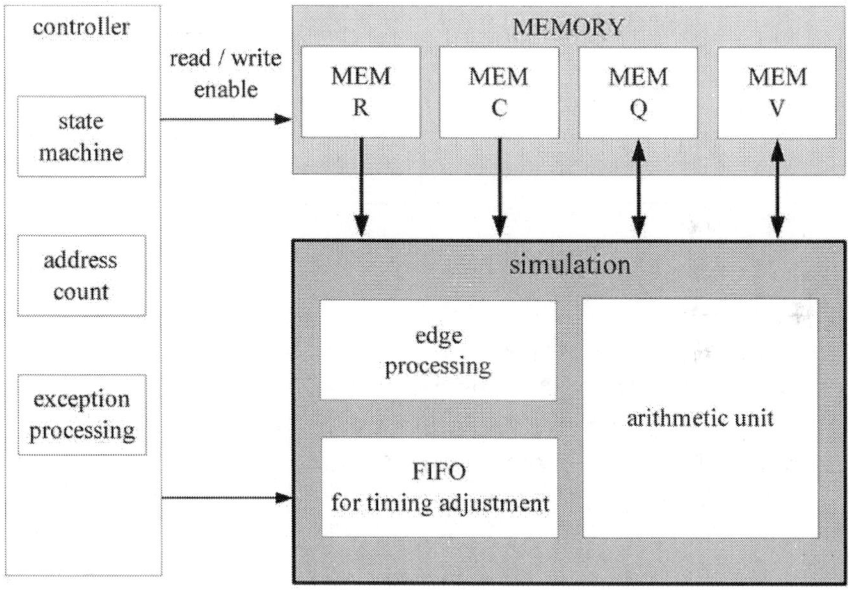

Figure 7. Block diagram of the power grid simulation.

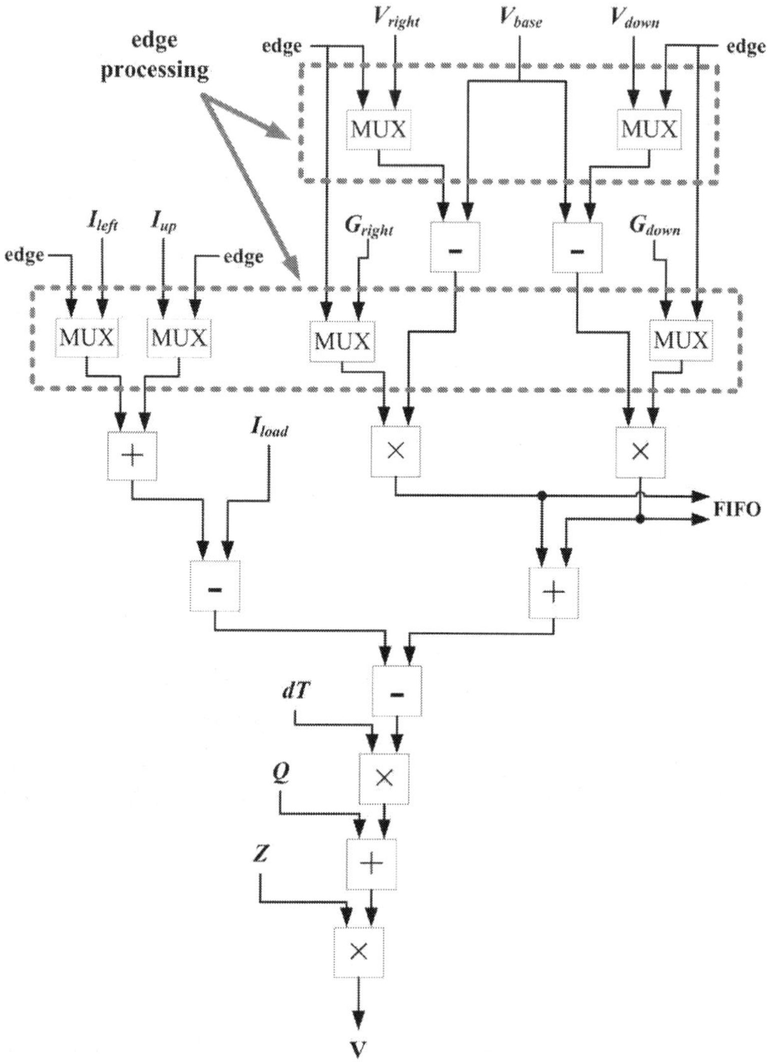

Figure 8. Calculation procedure in the simulation module.

The simulation module processes the circuit variables from memory, and writes simulation result into the memory. The variables written in the memory are the current, the charge, and the voltage. The current is used to calculate the RISK, the charge is necessary for each grid simulation, and the voltage is both. When the edge of the power grid is simulated, the variable is switched to the exception parameter by multiplexer (MUX) because the adjacent SLOT doesn't exist.

HARDWARE ALGORITHM

Pipeline Processing

Registers are inserting between each arithmetic unit to achieve pipeline processing. The proposed pipeline processing is composed of eight stages, and the data of each SLOT is transferred to the simulation module per clock cycle. The simulation flow by pipeline processing is as follows.

1) Each variable is stored from the memory to register.
2) Calculation of potential difference.
3) Calculation of each current.
4) Addition of each current (I_{right}, I_{down}, I_{left} and I_{up}).
5) Addition of Stage 4 and current source.
6) Calculation of inflow/outflow currents in small time step.
7) Computation of charge.
8) Computation of voltage.

Figure 9 shows the pipeline stage in the simulation module. All nodes of the power grid are sequentially simulated in the simulation module.
Because current calculation of a node require the voltage of the adjacent node, it is necessary to wait until the adjacent node finishes simulating, even if the node finishes simulating. Horizontal size and vertical size of power grid are defined as H_SIZE and V_SIZE. In an ideal pipeline processing, H_SIZE times V_SIZE clocks are needed for the simulation of all nodes. Moreover, 8 + H_SIZE clocks are needed to finish eight stage pipeline processing of a node and the adjacent node.

Parallel Processing

Figure 10 shows an example of which a circuit is divided into N parts to perform parallel processing. The memory and the simulation module are regarded as one functional block, and all the functional blocks are operated in parallel. The boundary of sub-circuit is overlapped, and the data of boundary part is stored in each memory.

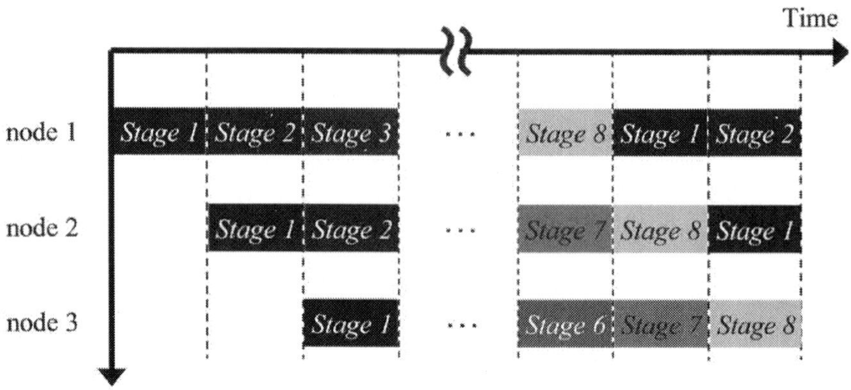

Figure 9. Pipeline stage of each node.

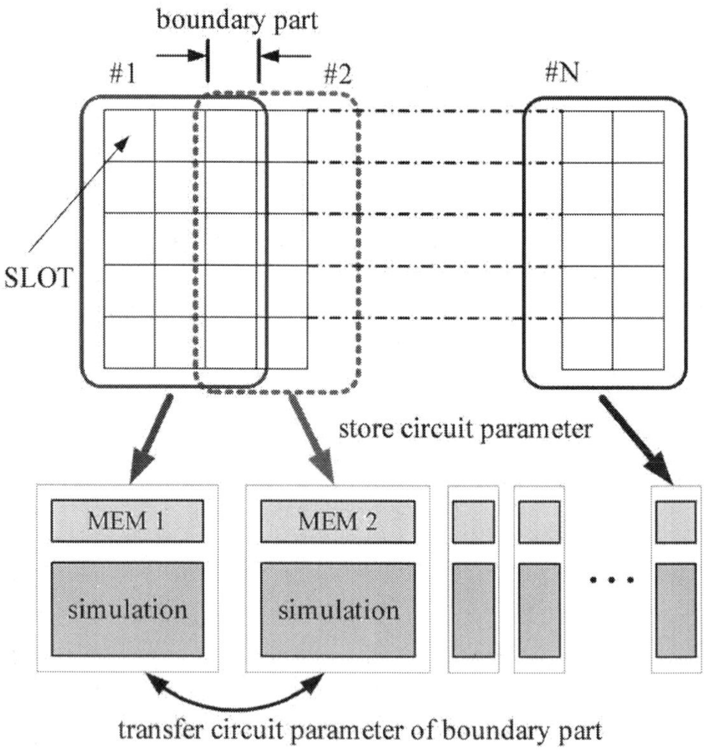

Figure 10. Circuit partitioning for parallel processing.

There is a point that should be considered in parallelization. For example, when simulating the sub-circuit No. 2, voltage is referred from right-hand memory to compute the outflow current of right boundary part. The inflow current is considered to simulate the left boundary part, however the inflow current quotes the outflow current of the left node as shown in **Figure 4**. However, referring to the inflow current, it is necessary to adjust timing at Stage 2. Then, simulation results are referred from left-hand circuit, and the process of timing adjustment is omitted to prevent performance deterioration. Therefore, the results of boundary part must be stored into each memory.

In massively parallel computing approach, e.g. GPGPU, similar synchronization technique is discussed to hold the important data in shared memories. In our FPGA approach, the structure of the shared memories are structured as we like, thus, in general, more efficient data communications are possible.

EXPERIMENTAL RESULTS

Evaluation of Pipeline Processing

To evaluate the speed of the proposed hardware simulator, FE is described in C language with fixed point, which is functionally equal to the hardware simulator. Bit shift is executed to prevent overflow after multiply operation. The simulation is executed with fixed time step. The FE program in C language is executed on HP xw9400Workstation with Dual-Core AMD Opteron Processor2220 2.8 GHz and 8 GB RAM. The test data for accuracy evaluation is given by an RC circuit as shown in**Figure 1**, and the circuit scale is 100 × 100 grids. The value of resistance is selected from 2, 3, 4 [mΩ], and the value of each capacitance is selected from 10, 30, 60 [μF].

The supply voltage is 1 [V]. The results of SPICE simulation by the same circuits are used for the reference data. SPICE performs Backward Euler method and Trapezoidal method [14]. These analytical methods realize high accuracy, however take a lot of processing time, in general. For speeding up the simulation, SPICE dynamically selects the time step size. The maximum error ratio in the all node voltage is evaluated by comparing the FE and SPICE.

First, pipeline processing is achieved as one of the speed-up techniques with hardware. The architecture described in Section 4 is achieved with Verilog HDL. Then, the speed performance of the power grid simulation by three processing, software, non-pipeline, pipeline, are compared as shown in **Table 3**. The processing time of software shows the result of single thread execution. That of HDL is calculated by number of clock times multiplied by the clock period of the FPGA. Target FPGA for logic synthesis is “EP2S180F1020C3, Stratix II, ALTERA”. Quartus II 7.1 is used as a logic synthesis tool. Non-pipeline processing is achieved about 1.2 times faster than software. As a result, the maximum frequency, 26 [MHz], was low because some operations had been executed with one clock. Inserting the register between each arithmetic unit, pipeline processing achieved about three times faster than non-pipeline processing. The maximum clock frequency and the latency are 80 [MHz] and 101 [nsec] by pipeline processing. In this experiment, 32-bit multiplication has become critical path because the pipeline stage was delimited between each arithmetic unit. About dividing the pipeline stage, there is leeway for improvement.

Next, the circuit scale has been changed. **Table 4** shows the speed gain by the hardware implementation. Experiment has been executed on large power grid to compare simulation time. The speed gain tends to be high in large scale circuit, and the hardware simulator has achieved 4.5 times faster processing speed than software.

Table 3. Speed gain by pipeline processing.

	Time [sec]	Frequency [MHz]
C	8.0	[N/A]
Non-pipeline	6.4	26
Pipeline	2.1	80

Table 4. Speed estimation by changing circuit scale.

Circuit scale	C [sec]	HDL [sec]
100 × 100	8.0	2.1
200 × 200	38	8.3
300 × 300	85	18
500 × 500	252	52

Evaluation of Parallel Processing

This section described parallel processing of the power grid simulation. The simulation module by pipeline processing in previous section is connected in parallel. The entire circuit scale is changed from 50 × 50 to 500 × 500, and the speeding gain by the parallel processing is shown in **Table 5**.**Figure 11** shows the processing time and speed ratio when the circuit scale is set to 500 × 500 grids. The bar chart indicates the processing time. The line charts and shows the speeding up ratio when the number of partitions is expanded.

The speeding up ratio is set on the basis of the processing time at one block. A high parallelism have achieved without saturation even if the number of partitions increased. About 7.9 times speeding up has been achieved when the power grid had been divided into eight. When the number of division circuits is defined as N, the speed improvement of proposal algorithm expect as shown in (1).

$$1+(N-1)\times(\mathrm{H\_SIZE}-1)/\mathrm{H\_SIZE} \tag{1}$$

Increasing the number of division circuit influence speed improvement because the boundary parts are overlapped. **Table 6** shows the result of logic synthesis when simulating 100 × 100 grids and divided into eight. The usage rate of DSP is comparatively high. Therefore, it is necessary for higher parallel processing to improve the algorithm or apply to larger scale FPGA.

CONCLUSIONS

In this paper, we have proposed an efficient hardware circuit simulator for power grid optimization. The proposal technique achieves high accuracy and high speed simulation by adopting fixed point arithmetic and parallel processing.

In the evaluation experiment of accuracy, we have evaluated four type's analysis method by comparing the functionally equal program and SPICE. The FE achieves high accuracy simulation by several different experiments, i.e., in different time steps, floating point or fixed point.

Table 5. Speed gain by changing the number of partitions.

	2	3	4	5	6	7	8
50 × 50	2.0	2.9	3.8	4.6	5.5	6.3	7.1
100 × 100	2.0	2.9	3.8	4.8	5.6	6.2	7.1
200 × 200	2.0	3.0	3.9	4.9	5.9	6.8	7.7
500 × 500	2.0	3.0	4.0	5.0	5.9	6.9	7.9

Table 6. Result of logic synthesis of eight parallel simulation modules.

Logic	14,546 [LEs] (10%)
Memory	4.3 [Mbit] (45%)
DSP (9-bit)	524 (68%)
Max frequency.	80 [MHz]

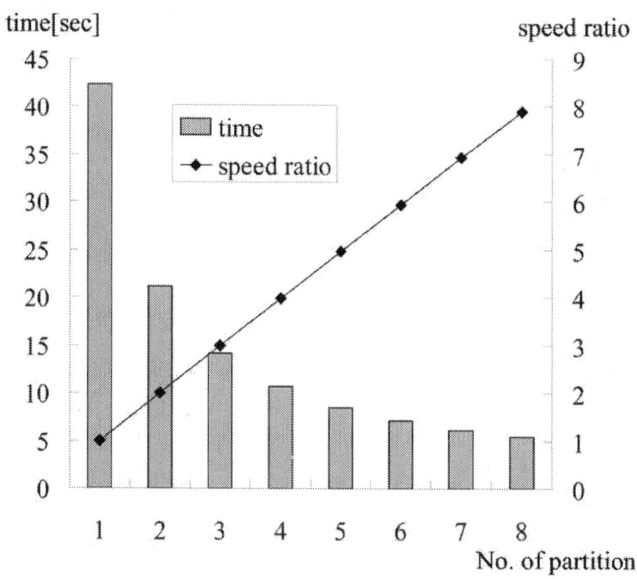

Figure 11. Circuit partitioning for parallel processing.

Next, we have evaluated the speed gain by pipeline processing and parallel processing. The proposed power grid simulation algorithm performs 4.5 times faster processing than software processing. In addition, eight parallel processing achieves 7.9 times higher speed than one unit processing. Therefore, the proposed power grid simulation using 80 MHz FPGA achieves 35 times higher speed than software processing with 2.8 GHz CPU while maintaining the high accuracy.

In the future, we will implement the proposed hardware algorithm onto a Compute Unified Device Architecture (CUDA) platform.

REFERENCES

1. D. A. Andersson, L. J. Svensson and P. Lasson-Edefors, "Noise-Aware On-Chip Power Grid Considerations Using a Statistical Approach," Proceedings of International Symposium on Quality Electronic Design, San Jose, 17-19 March 2008, pp. 663-669.
2. S. W. Wu and Y. W. Chang, "Efficient Power/Ground Network Analysis for Power Integrity-Driven Design Methodology," Proceedings of Design Automation Conference, San Diego, 7-11 June 2004, pp. 177-180.
3. A. Muramatsu, M. Hashimoto and H. Onodera, "Effects of On-Chip Inductance on Power Distribution Grid," IEICE Transactions on Fundamentals of Electronics, Communications and Computer Sciences, Vol. E88-A, No. 12, 2005, pp. 3564-3572.
4. Y. Zhong and M. D. F. Wong, "Thermal-Aware IR Drop Analysis in Large Power Grid," Proceedings of International Symposium on Quality Electronic Design, San Jose, 17-19 March 2008, pp. 194-199.
5. B. Yu and M. L. Bushnell, "Power Grid Analysis of Dynamic Power Cutoff Technology," Proceedings of International Symposium on Circuits and Systems, New Orleans, 27-30 May 2007, pp. 1393-1396.
6. C. Mizuta, J. Iwai, K. Machida, T. Kage and H. Matsuda, "Large-Scale Linear Circuit Simulation with an Inversed Inductance Matrix," Proceedings of Asia and South Pacific Design Automation Conference, Kanagawa, 27-30 January 2004, pp. 511-516.
7. P. M. Lee, S. Ito, T. Hashimoto, J. Sato, T. Touma and G. Yokomizo, "A Parallel and Accelerated Circuit Simulator with Precise Accuracy," Proceedings of International Conference on VLSI Design, Bangalore, 7-11 August 2002, pp. 213-218.
8. N. Nakasato and T. Hamada, "Acceleration of Hydrosynamical Simulations Using a FPGA Board," Institute of Electronics, Information and Communication Engineers Technical Report, Vol. 105, No. 515, 2006, pp. 19-24.

9. T. Watanabe, Y. Tanji, H. Kubota and H. Asai, "Parallel-Distributed Time-Domain Circuit Simulation of Power Distribution Networks with Frequency-Dependent Parameters," Proceedings of Asia and South Pacific Conference on Design Automation, Yokohama, 24-27 January 2006, pp. 832-837.
10. Y. Gu, T. Vancourt and M. C. Herbordt, "Improved Interpolation and System Integration for FPGA-Based Molecular Dynamics Simulations," Proceedings of International Conference of Field Programmable Logic and Applications, Madrid, 28-30 August 2006, pp. 1-8.doi:10.1109/FPL.2006.311190
11. L. Zhuo and V. K. Prasanna, "High-Performance and Parameterized Matrix Factorization on FPGAs," Proceedings of International Conference of Field Programmable Logic and Applications, Madrid, 28-30 August 2006, pp. 363-368. doi:10.1109/FPL.2006.311238
12. M. Yoshimi, Y. Osana, Y. Iwaoka, Y. Nishikawa, T. Kojima, A. Funahashi, N. Hiroi, Y. Shibata, N. Iwanaga, H. Kitano and H. Amano, "An FPGA Implementation of Throughput Stochastic Simulator for Large-scale Biochemical Systems," Proceedings of International Conference of Field Programmable Logic and Applications, Madrid, 28-30 August 2006, pp. 227-232. doi:10.1109/FPL.2006.311218
13. H. Ishijima, T. Harada, K. Kusano, M. Fukui, M. Yoshikawa and H. Terai, "A Power Grid Optimization Algorithm with Consideration of Dynamic Circuit Operations," Proceedings of Synthesis and System Integration of Mixed Information, Nagoya, 3-4 April 2006, pp. 446- 451.
14. Y. Kawakami, M. Terao, M. Fukui and S. Tsukiyama, "A Power Grid Optimization Algorithm by Observing Timing Error Risk by IR Drop," IEICE Transactions on Fundamentals, Vol. E91-A, No. 12, 2008, pp. 3423-3430.
15. T. Hashizume, H. Ishijima and M. Fukui, "An Evaluation of Circuit Simulation Algorithms for Hardware Implementation," Proceedings of Synthesis and System Integration of Mixed Information, Hokkaido, 15-16 October 2007, pp. 322-327.

CITATION

T. Hashizume, M. Yoshikawa and M. Fukui, "A Parallel Circuit Simulator for Iterative Power Grids Optimization System," *Circuits and Systems*, Vol. 3 No. 2, 2012, pp. 153-160. doi: 10.4236/cs.2012.32020.

Chapter 8

A Novel High-Performance LEKAGE-Tolerant, Wide Fan-In Domino Logic Circuit in Deep-Submicron Technology

Ajay Dadoria, Kavita Khare, T. K. Gupta, R. P. Singh

Electronics & Communication Engineering, MANIT, Bhopal, India

ABSTRACT

As technology shrinks in modern era the demand on high speed, low power consumption and small chip area in microprocessors is come into existence. In this paper we have presented a new class of domino circuit design for low power consumption, faster circuit speed and high performance. Due to wide fan-in domino logic, its logic gate suffer from noise sensitivity, if we improve sensitivity, sub-threshold and gate oxide leakage current dominate in evaluation network, which increases the power consumption and reduces the performance of the circuit. The proposed circuit improves the dynamic power consumption and reduces the delay which improves the speed of the circuit. Simulation is performed in BISM4 Cadence environment at 65 nm process technology, with supply voltage 1 V at 100 MHz frequency and bottleneck operating temperature of 27˚C with C_L= 1 fF. From the result average power improvement by proposed circuit 1 & 2 for 8 input OR gate is 10.1%, 15.28% SFLD, 48.56%, 51.49% CKD, 55.17%, 57.71% HSD and improvement of delay is 1.10%, 12.76% SFLD, 19.13%, 28.63% CKD, 4.32%, 15.59% HSD, 19.138%, 44.25% DFD respectively.

INTRODUCTION

As technology scales down power consumption is dominant in deep sub-micron technology. Power consumption is divided into two parts static and dynamic circuit. When we talk about low power, high speed, we prefer dynamic logic instate of static CMOS circuits, by generating proper logic at output of the circuit. All these properties of the dynamic node make it high robust circuit, to generate proper logic in dynamic CMOS circuits, so to achieve high performance. Main limitations of dynamic logic occur during cascading of similar circuit during cascading of large circuit an erroneous state occurs in dynamic logic design; another problem which come into existence is charge sharing which reduces the voltage of dynamic node. So to overcome from the problem a week PMOS transistor (with a small (W/L) ratio) placed parallel to pre-charge transistor, which is feedback from the output, the output of the circuit is connected with inverter for reducing the cascading problem [1] .

Domino logic circuit is also a kind of dynamic logic circuit which is used for the high speed and high performance application. Also the domino logic circuit plays a vital role where fan in are high in any circuit. Domino circuits are widely used in high performance microprocessors, register files, ALU, DSP circuits and priority encoders in content addressable memories, such as high fan-in multiplexer or comparator circuits. A basic footer less domino logic circuit is shown in Figure 1 where in pre-charge phase (clk = 0) PMOS transistor ON and charge the dynamic node, during evaluation phase (clk = 1) the dynamic node discharge when any one of the input is ON, the output of the inverter is feed back to week keeper transistor so to maintain the charge on dynamic node.

In this paper, a new domino circuit for wide fan-in applications in ultra deep sub-micrometer technologies is proposed. The novelty of the proposed circuit is that our work simultaneously reduces the delay and leakage power consumption.

The rest of the paper is organized as follows. Section 1 explains some basics of domino logic circuits. Typical conventional approach has been discussed in Section 2. Section 3 gives the brief of proposed circuit for the betterment of the noise margin. Simulation result and conclusion are discussed in Section 4 and Section 5 respectively.

LITERATURE REVIEW

The most popular domino logic is the standard footless domino [2] circuit as shown in Figure 1. For enhance the performance of the microprocessor, modification is done on circuit level to increase the robustness of the circuit, without penalty of noise immunity, the keeper transistor is added. Keeper transistor is feedback from the output; keeper transistor W/L ratio is very low to maintain the charge in the dynamic node.

The keeper ratio K is defined as

$$K = \frac{\mu p\left(\frac{W}{L}\right)\text{Keeper} - \text{transistor}}{\mu n\left(\frac{W}{L}\right)\text{evaluation} - \text{network}} \qquad (1)$$

where W and L denote the transistor size, and mobility of electron and hole is represented by μn and μp. However by inserting the keeper transistor performance of the circuit is degraded and power dissipation increases. Upsizing the keeper transistor is another solution for improves robustness but it result in higher power dissipation and delay [3] . Contention between evaluation network and keeper transistor, therefore size of the keeper transistor should be low to achieve high-speed in microprocessor. Thus, trade off exist between delay and power to improve noise and leakage immunity [4] . Several techniques introduce in this section to address this issue.

High Speed Domino Logic (HS)

High speed domino is another domino logic circuit. In domino logic circuit current drawn through the keeper transistor and pull down network NMOS transistors at the beginning of the evaluation phase, can be reduced by applying a clock delay in the circuit. That does not affect the leakage current in the circuit [5] . But apart from this the extra clock delay consumes extra area and power, which is a big drawback of the circuit as shown in Figure 2. In High speed domino logic circuit when clock becomes high, M_{n1} is still off and M_{p2} is still on. Therefore M_{p2} turns off the keeper transistor. After some delay of inverter M_{p2} becomes off. Hence at the beginning of evaluation phase dynamic node is afloat, so in the absence of keeper transistor, evaluation node may be discharged for any noise at the input section [7] . Also the voltage at the gate of the keeper transistor would be V_{DD}-V_{tMn1}. This would provide a dc current flow through the PMOS keeper transistor and the NMOS network.

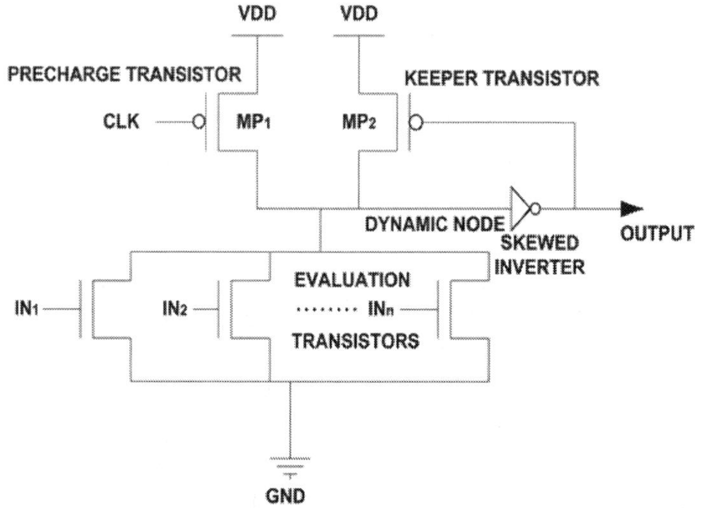

Figure 1. Standard footer less domino logic.

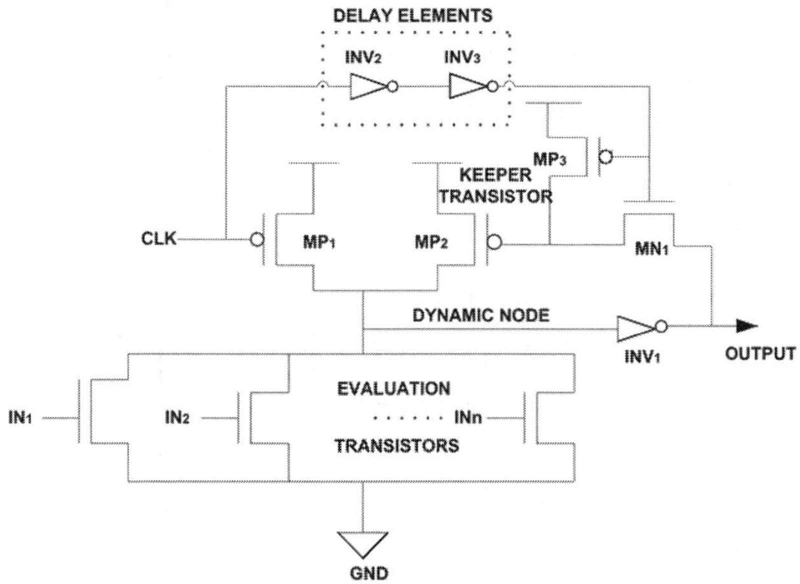

Figure 2. High speed domino logic.

Conditional Keeper Domino Logic (CKD)

The conditional keeper domino logic contains two PMOS transistor keeper circuit [6] . In which one is of smaller strength and other is of higher strength as shown in Figure 3. Now when the dynamic node is at high voltage M_{kp1} gets turn on to avoid voltage drop at the dynamic node. If the dynamic node is still high, then after a certain amount of delay, during the evaluation phase output of NAND gate becomes low which makes M_{k2} to turn on? For maintaining the state of dynamic node, is M_{k1}responsible during the beginning of evaluation phase and M_{k2} is responsible for the rest of the evaluation phase [7] .

Diode Footed Domino (DFD)

In diode footed domino we modify the conventional domino circuit by adding an nMOS transistor M_1in series with the foot of the evaluation network. This nMOS transistor is in diode configuration i.e. gate and drain terminals connected together. Figure 4 [9] shows the Diode Footed Domino configuration. Stacking effect [10] occur because this transistor M_1 is connected in series with the evaluation network [12] [13] . Thus sub-threshold leakage current reduces as a result of stacking effect. DFD circuit works as follow:

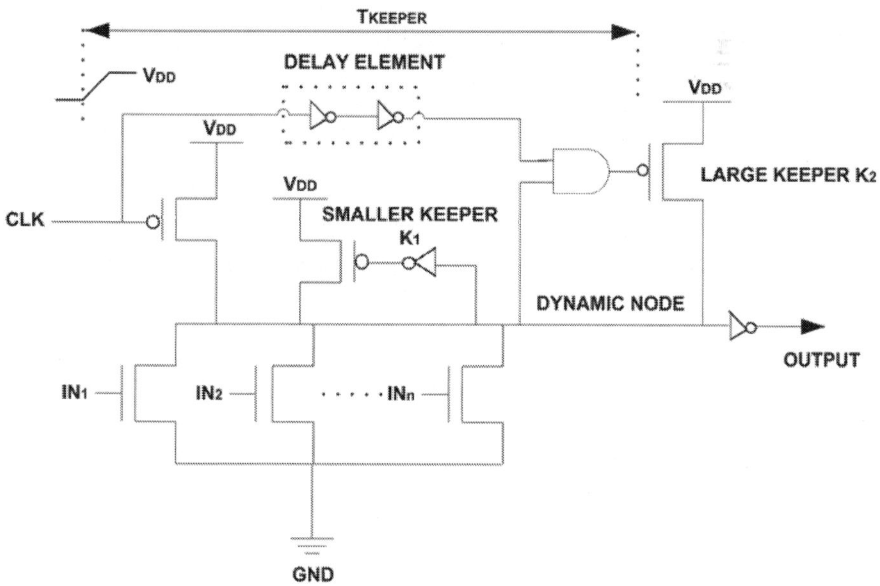

Figure 3. Conditional keeper domino logic.

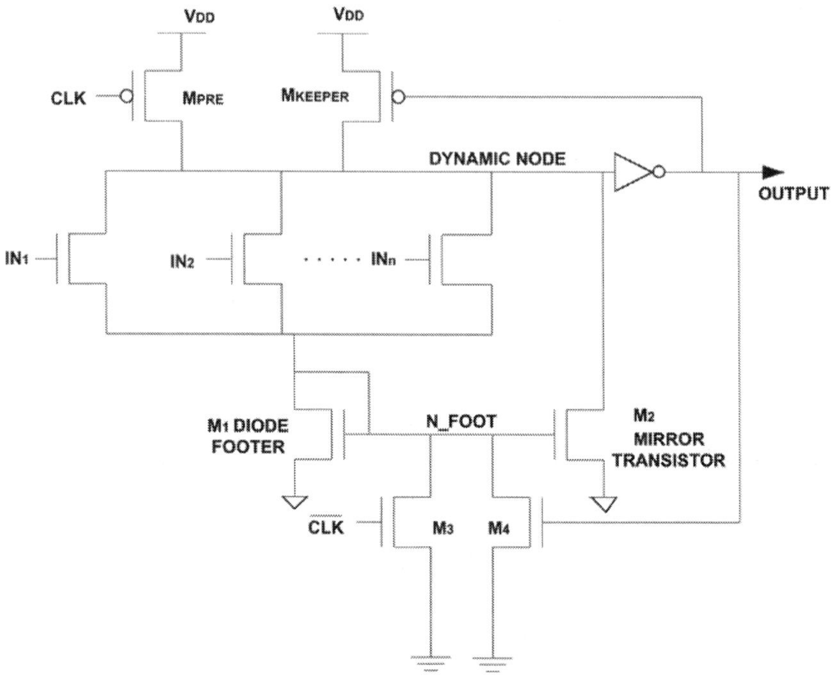

Figure 4. Diode footed domino logic.

The mirror ratio is defined as the ratio of the current drivability of the mirror transistor to that of the diode footer:

$$M = \frac{\left(\frac{W}{L}\right)_{\text{mirror transistor}}}{\left(\frac{W}{L}\right)_{\text{diode footer}}}$$

By increasing the mirror ratio, the performance can be increased.

PROPOSED CIRCUIT

In conventional footer less circuit, during precharge phase (clk = 0), PMOS transistor ON and charge the dynamic node from V_{dd}, during evaluation phase (clk = 1), a dynamic node not able to maintain the constant because PMOS transistor rail OFF from $V_{dd,}$ only keeper Transistor connected to V_{dd} maintain the charge of dynamic node if all the transistor is OFF in evaluation network as shown in Figure 5. During evaluation phase when any one input is ON of NMOS block the dynamic node will discharge, which result in flow of sub-thre- shold and gate oxide leakage current which result in degradation of UNG of the circuit, for reduction of leakage current and enhance the noise immunity of the circuit we have proposed a circuit.

In proposed circuit modification is done in evaluation network, we have inserted two NMOS transistor between dynamic node and pull down network. To improve the efficiency of the proposed circuit and extra NMOS transistor is connected to the dynamic node to produce the proper stacking of the evaluation network, to increase the noise immunity of the circuit and reduces the leakage current of the circuit by providing half swing logic at the output node as sown in Figure 6. In footed portion we place NMOS transistor, during pre-charge phase footed transistor is OFF, during evaluation phase a charge discharge from dynamic node the two NMOS transistor provides the stacking effect for leakage reduction and high noise immunity. In proposed circuit 2 as shown inFigure 7. Input of footed transistor is connected to clock and three inverter for generation of delay in the footed transistor, the delay element is used for proper slower the gate and greater noise robustness. These approaches do not reduce the overall leakage current, but only the leakage current at the dynamic node that drives the final static inverter and is the critical node. Hence we have more degree of freedom for increasing speed or enhance noise immunity by reducing the leakage current.

SIMULATION RESULTS AND COMPARISON

The proposed circuit was simulated using BISM4 Cadence virtuoso schematic editor tool in the high-perfor- mance 65-nm predictive technology and at the temperature of 27°C. The supply voltage used in the simulations is 1 V, for 8, 16 and 32 input OR gates as shown in Tables 1-4. Various parameters have been mitigated such as power dissipation, delay and UNG (Unit Noise Gain). Table 4 calculate the standby leakage current at different temperature for 8, 16 and 32 inputs, operating frequency is 100 MHz with C_L = 1 pF of the circuit. Average power consumption of existing domino logic circuit and proposed circuit is shown in Figure 8. UNG graph is shown in Figure 9.

The noise-margin metric used in this paper is called the unity noise gain (UNG). For calculation of UNG [11] , a pulse noise is applied to all inputs with amplitude which is a fraction of supply voltage and a pulse width equal to 30% of duty cycle. This noise amplitude is defined as

$$UNG = \{Vin, Vnoise = Voutput\}$$

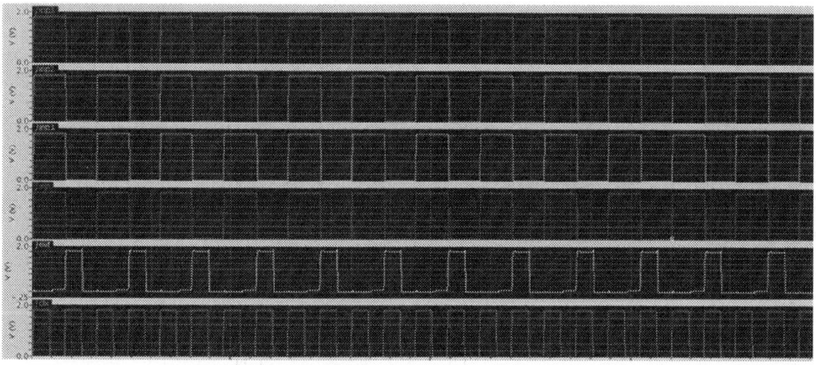

Figure 5. Output wave form of proposed circuit.

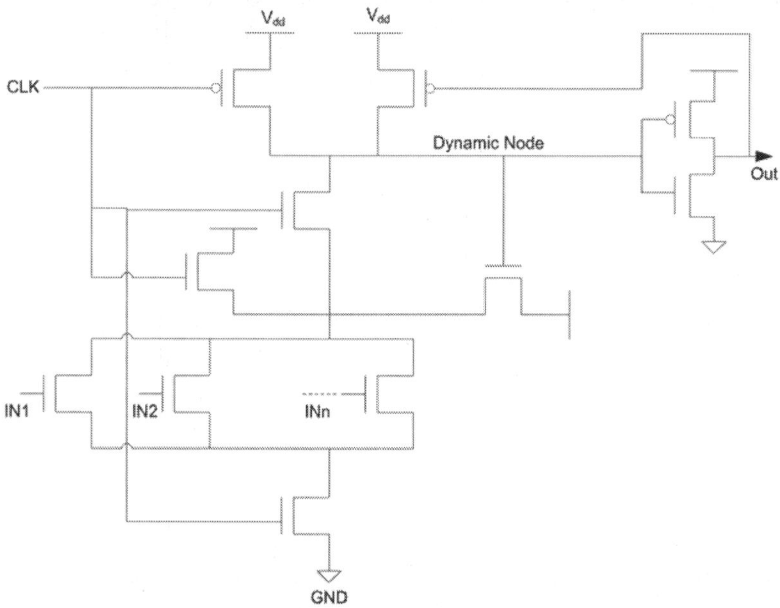

Figure 6. Proposed circuit 1.

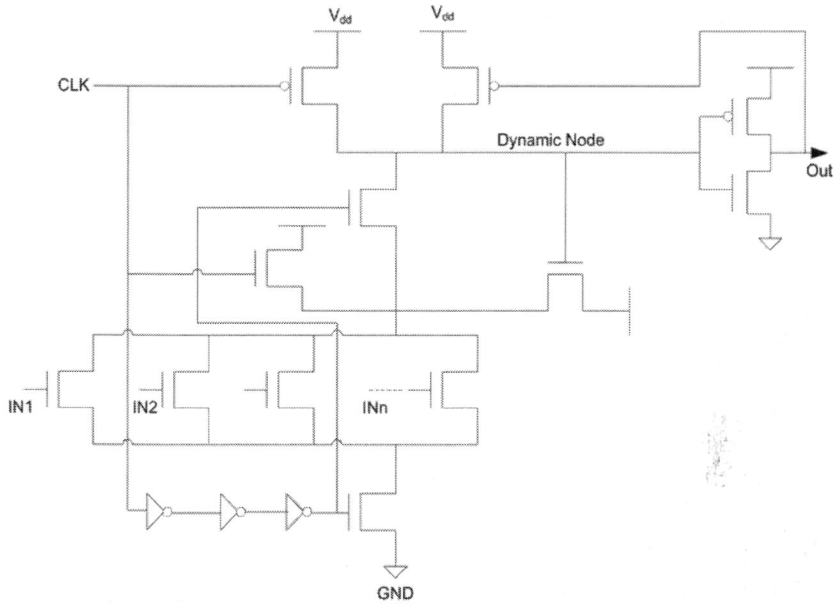

Figure 7. Proposed circuit 2.

Table 1. Comparison of power dissipation (μW).

S. No.	Logic Style	8 Input	16 Input	32 Input
1.	SFLD	2.611	4.234	7.632
2.	CKD	4.56	6.323	9.862
3.	HSD	5.231	8.214	10.12
4.	DFD	1.672	2.540	3.965
5.	Proposed Circuit 1	2.345	5.203	8.812
6.	Proposed Circuit 2	2.212	4.780	7.642

Table 2. Comparison of UNG (in Volt).

S. No.	Logic Style	8 Input	16 Input	32 Input
1.	SFLD	0.252	0.228	0.212
2.	CKD	0.264	0.225	0.199
3.	HSD	0.243	0.230	0.203
4.	DFD	0.274	0.256	0.2574
5.	Proposed Circuit 1	0.263	0.231	0.203
6.	Proposed Circuit 2	0.254	0.224	0.191

Table 3. Comparison of Delay (in ps).

S. No.	Logic Style	8 Input	16 Input	32 Input
1.	SFLD	13.102	18.435	28.59
2.	CKD	16.015	21.2	31.23
3.	HSD	13.541	18.32	28.43
4.	DFD	16.49	22.16	33.12
5.	Proposed Circuit 1	12.95	15.651	21.34
6.	Proposed Circuit 2	11.432	14.125	19.57

Table 4. Stand by leakage current of proposed circuit (in nA).

S. No.	Temperature	8 Input	16 Input	32 Input
1.	27°C	262.25	308.79	401.11
2.	50°C	364.12	411.34	505.45
3.	110°C	700.24	754.23	851.34

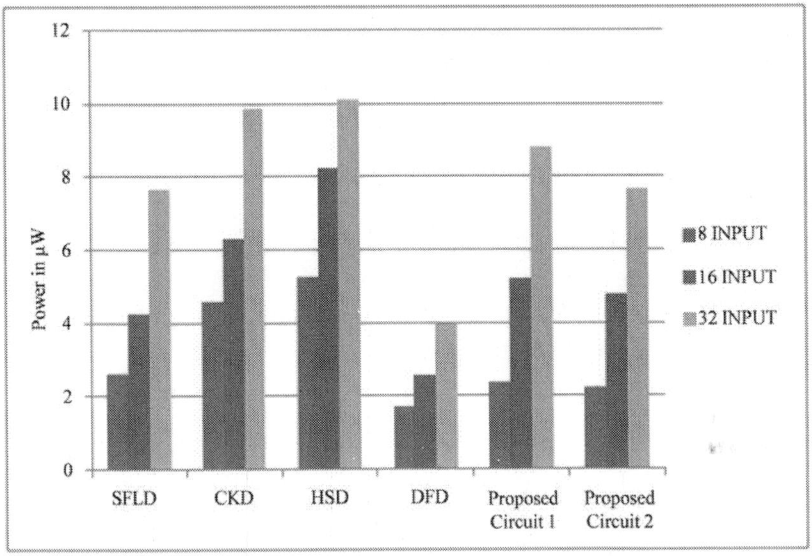

Figure 8. Comparison of average power.

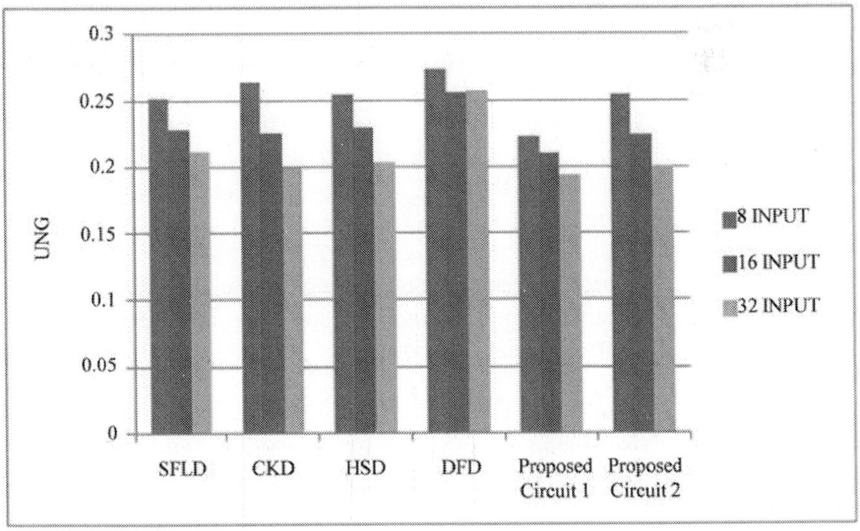

Figure 9. Comparison of UNG.

CONCLUSION

Domino logic dissipates very low standby power compare to static CMOS logic. The novelty of the proposed circuit is that domino circuit reduces the leakage power consumption by maintaining the same level of delay and UNG. In this paper our main objective was to improve the noise immunity and to reduce the average power consumed and delay associated with the circuit. The analysis shows that we get an appreciable improvement in the noise margin and power consumption and marginal improvement in delay. The whole comparison is based upon 65 nm CMOS technology using Cadence Virtuoso tool.

REFERENCES

1. Clarke, L.T. and Taylor, G.F. (1996) High Fan-In Circuit Design. IEEE Journal of Solid-State Circuits, 31, 91-96. http://dx.doi.org/10.1109/4.485870
2. Tsui, B.-Y. and Chin, L.-F. (2004) A Comprehensive Study of the FIBL of Nanoscale MOSFETs. IEEE Transactions on Electron Devices, 51, 1733-1735.
3. Moradi, F., VuCao, T., Vatajelu, E.I., Peiravi, A., Mahmoodi, H. and Wisland, D.T. (2012) Domino Logic Designs for High-Performance and Leakage-Tolerant Applications. Elsevier INTEGRATION, the VLSI Journal, 46, 247-254.
4. Moradi, F., Peiravi, A. and Mahmoodi, H. (2004) A New Leakage Tolerant Design for High Fan-in Domino Circuits. IEEE Conference, 493-496.
5. Allam, M.W., Anis, M.H. and Elmasry, M.I. (2000) High-Speed Dynamic Logic Style for Scaled-Down CMOS and MTCMOS Technologies. Proceedings of the International Symposium on Low Power Electronics and Design, 155-160.http://dx.doi.org/10.1109/LPE.2000.876774
6. Alvandpour, A., Krishnamurthy, R.K., Soumyanath, K. and Borkar, S.Y. (2002) A Sub-130-nm Conditional Keeper Technique. IEEE Journal of Solid-State Circuits, 37, 633-638.http://dx.doi.org/10.1109/4.997857
7. Alvandpour, A., Krishnamurthy, R., Soumayanath, K. and Borkar, S. (2001) A Low-Leakage Dynamic Multi Ported Register File in 0.13 μm CMOS. Proceedings of International Symposium on Low Power Electronics and Design, 68- 71.
8. Peiravi, A. and Elissa, M.A. (2013) Current-Comparison-Based Domino: New Low-Leakage High-Speed Domino Circuit for Wide Fan-In Gates. IEEE Transactions on Very Large Scale Integration (VLSI) Systems, 21.
9. Lih, Y., Tzartzanis, N. and Walker, W.W. (2007) A Leakage Current Replica Keeper for Dynamic Circuits. IEEE Journal of Solid-State Circuits, 42, 48-55.http://dx.doi.org/10.1109/JSSC.2006.885051
10. Yorozu, Y., Hirano, M., Oka, K. and Tagawa, Y. (1987) Electron Spectroscopy Studies on Magneto-Optical Media and Plastic Substrate Interface. IEEE Translation Journal on Magnetics in Japan, 2, 740-741. [Digests 9th Annual Conf. Magnetics Japan, p. 301, 1982].

11. Peiravi, A., Moradi, F. and Wisland, D.T. (2009) Leakage Tolerant, Noise Immune Domino Logic for Circuit Design in the Ultra Deep Submicron CMOS Technology for High Fan-In Gates. Journal of Applied Sciences, 9, 393-396.http://dx.doi.org/10.3923/jas.2009.392.396
12. Mahmoodi-Meimand, H. and Roy, K. (2004) Diode-Footed Domino: A Leakage-Tolerant High Fan-in Dynamic Circuit Design Style. IEEE Transactions on Circuits and Systems—I, 51, 495-503.
13. Gupta, T.K. and Khare, K. (2014) A New Technique for Leakage Reduction in 65 nm Footerless Domino Circuits. Scientific Research Circuits and Systems, 4.

CITATION

Dadoria, A. , Khare, K. , Gupta, T. and Singh, R. (2015) A Novel High-Performance Lekage-Tolerant, Wide Fan-In Domino Logic Circuit in Deep-Submicron Technology. *Circuits and Systems*, **6**, 103-111. doi:10.4236/cs.2015.64011.

Chapter 9

A Novel Integrated Circuit Driver for Led Lighting

Yanfeng Jiang, Dong Zhang, Ming Yu

Beijing Institute of Auto-Testing Technology, Beijing, China

ABSTRACT

A novel integrated circuit for driving LED lighting has been proposed, designed and fabricated. Besides the typical parts of LED driver, an integral part was added at the output terminal of error amplifier in the driver. In this way, a novel average current mode can be set up to take the place ordinary peak current control mode. In addition, a BUCK low-level topology was adopted, too. It can be used to drive up to eight 1 W HB LED lights with 350 mA constant current. In this way, the LED driver displays high performance, in which output current with less 1% error and total efficiency as high as 96%. The feasibility of the design has been verified by actual measurement on the fabricated chip.

INTRODUCTION

Contrasting to the traditional fluorescent lamp, high bright LED, being a potential lighting source, has the obvious advantage in energy saving, environmental protection, high efficiency, etc. [1] . So it has attracted much attention in academy and industry to push forward its utility. Based on the requirements of HB LED, some detailed limitations should be considered during the design of LED driver. For example, the life of LED will be dramatically decreased on the condition that there exists the driving current deviation [1] . So an ideal LED driver should have the constant output current with the high accuracy [1] .

Currently, a peak current control mode has always been adopted in the existing LED driver to obtain constant current [2] . Based on statistical data [2] , 30% error could be caused by this peak current control mode, which subsequently deteriorates LED life time dramatically, up to 20%.

The main reason for influencing the accuracy of output current at peak current control mode lies in the fact that the peak current is varied with the inductor in the driver circuit [3] . Here, an average current control mode was adopted, in this case it isn't influenced by the inductor value anymore. In this way, a higher accurate output will be obtained in LED driver to guarantee its lifetime.

Besides the above mentioned accuracy problem, power efficiency is also a key parameter to evaluate the driver's quality. The topology of driver circuit will determine the efficiency [3] . In this paper, an optimized topology will be used to reduce power consumption.

To solve the above issues, an integrated circuit driver with average current control mode was proposed and designed in this paper, to get accurate and stable output current, which will be introduced in the Section 2. A BUCK low-level topology will be adopted in the driver to get high efficiency. This will be introduced in Section 3. Section 4 shows the simulation results using CSMC 0.5 μm 40 V BCD model. Finally, the experimental results and discussion will be made to show design's advantage.

CHIP DESIGN WITH AVERAGE CURRENT CONTROL MODE

Traditionally, in order to reduce the cost and complexity of the chip, a peak current mode control mode is always used in LED driver chip [4] [5] . However, the limitation of this method is the output current is easily influenced by inductor L and input voltage. This current variation will lead to degeneration of accuracy. At the same time, the actual current provided for the HB LED is less than the peak current. For this mode, theoretically speaking, the value of inductor should be infinity to obtain an output current which is equal to the design value.

Figure 1 schematically shows the current in the peak current control mode in the common BUCK topology. It can be seen from this figure that different inductor corresponding to same peak current, but different average currents. The greater the inductance L is, the closer the average approaches to peak current. Because of practical limitations, the inductance value L can not large enough. So the actual current value flowing through the LED is always less than the peak current. Based on the above analysis, there are two shortcomings for the peak current control mode. The one is that its average current is always varied by the inductor and input voltage. It can't maintain stable all the time. The other shortcoming lies in the fact that the average current is different from the designed peak current. These two facts have an obvious influence on the output's accuracy.

Figure 2 shows the schematic of the average current control mode. It can be seen that the average current always maintain the same regardless of inductor L values. Comparing the two control modes shown in Figure 1 and Figure 2, one conclusion can be made that the average current control mode can be an effective method to guarantee the accuracy of output current. So, in this work, the average current control mode will be fulfilled in the circuit design. An integrated circuit with this mode has been designed, which schematic is shown in Figure 3. The part within dashed line corresponds to the designed integrated circuit, which includes error amplifier (EA), pulse-width modulator (PWM), slope generator, RS flip-flop, output stage with power transistor, etc. Its structure is a typical PWM controller except there is an integral part at the output terminal of EA.

The integral part is composed by three components, c_1, c_2 and R_2. For the two input terminal of EA, positive one is connected to a voltage reference while negative one $V_{sense}(t)$, termed as the detection voltage, connected to outside sense resistor, as shown in Figure 3. When the power transistor is on-state, there will be a current flowing through the sense transistor, so the detection voltage will appear and moreover, it will be increased for the existence of an external inductor L, which is not displayed in this figure.

At beginning, detection voltage is smaller than reference, EA will have a positive output. Because the integral parts exist, capacitors c_1, c_2 will be charged so that the voltage at output terminal increases.

When the value of detection voltage is larger than the reference, output of EA should be negative. However, the integral part will try to maintain the original EA output voltage. So, there is a discharge phase during this period. There are two main work states: In the first state, when power transistor is on-state, the detection voltage will appear, increasing during the on-state, as shown inFigure 4(a). The output of EA, $V_{EA}(t)$, as shown in Figure 4(b), rises at the beginning. When the detection voltage equals to the reference, $V_{EA}(t)$ reaches to the maximum value. Afterwards, with detection voltage increasing, output of EA will be decreased.

In the second state, when N-MOSFET power transistor turns off, $V_{EA}(t)$ remains unchanged for the existence of integral circuit, as shown in Figure 4(b). Figure 4(c) shows the waveform of slope, which is connected to PWM's negative input terminal, denoted as $V_{slope}(t)$. $V_{EA}(t)$ is connected to positive ones. The average value of detection voltage during on-state has been set to be equal to the reference voltage. Figure 4(d) shows the output of PWM modulator. So the system is operated in a fixed period of conduction control mode. As is shown in the dotted line area in Figure 3, it is the core circuit for the average current regulation mode.

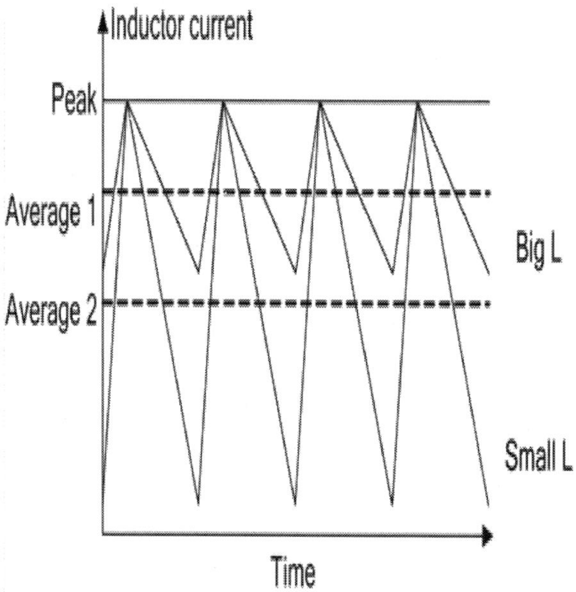

Figure 1. Schematic of peak current control mode.

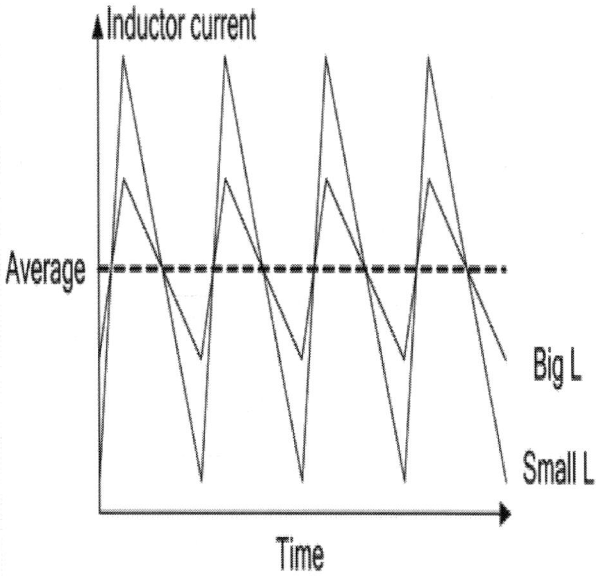

Figure 2. Schematic of average current control mode.

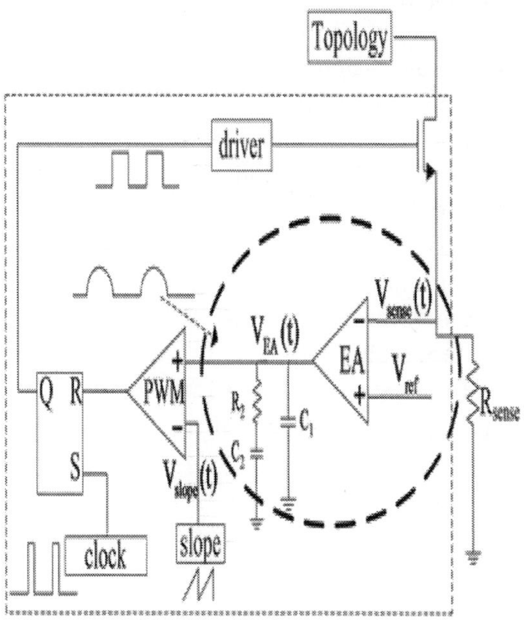

Figure 3. Schematic diagram of driving circuit.

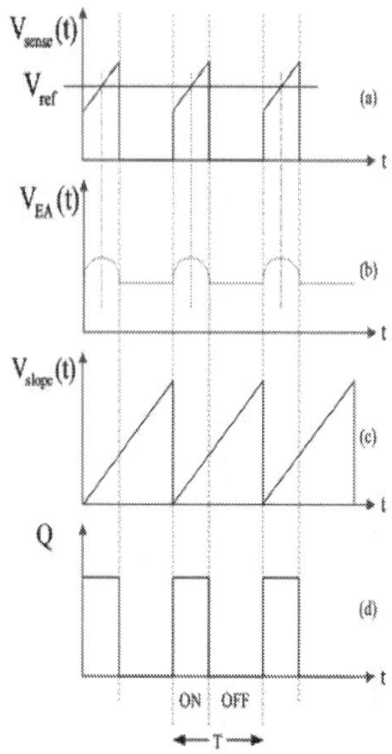

Figure 4. Timing diagram of driving circuit.

The traditional current regulator was designed as peak current control mode. It can't provide the actual average current regulation for the LED lamps. The traditional circuit uses an error amplifier to compare the difference between the transient value of detection voltage $V_{sense}(t)$ and the reference voltage. For the circuit shown in Figure 3, which is the average current mode and carries out the time integration about the error, the capacitors take the integration of the difference between the detection voltage $V_{sense}(t)$ and the reference voltage V_{ref}. Because the reference voltage V_{ref} is a constant value, that is to say, during the integration time $T*D$ (D is the switching duty cycle, T is the period time), it enables the difference between the average of $V_{sense}(t)$ and the reference voltage drop to a minimum value, so:

$$\int_0^{TD}\left[V_{ref}-V_{sense}(t)\right]dt=V_{ref}*TD-\int_0^{TD}V_{sense}(t)dt \tag{1}$$

Supposing Δv as the capacitance variation during Δt period, that is the increment of $V_{EA}(t)$. The equivalent output capacitance of EA is c, and the transconductance of EA is g_m. Following equation can be obtained:

$$\Delta V_{EA}(t)=\Delta v=\frac{i\Delta t}{c}$$
$$=\frac{\Delta t}{c}g_m\left[V_{\text{ref}}-V_{\text{sense}}(t)\right]$$
$$=\frac{g_m}{c}\left[V_{\text{ref}}-V_{\text{sense}}(t)\right]\Delta t \quad (2)$$

Based on Equation (2), when V_{ref} equals to $V_{sense}(t)$, the voltage on the capacitor reaches maximum value. When V_{ref} is less than $V_{sense}(t)$, the capacitor discharges, and then when V_{ref} is larger than $V_{sense}(t)$, the capacitor charges. So, during the TD period, $V_{EA}(t)$ increases at first. Then, it will decrease until the capacitor voltage equals to its initial value. When the state changes, the value of $V_{EA}(t)$ will remain the same during $T*(1-D)$ period. In this way, it repeats every cycle to form a stable average current.

It can also be noted that the output of EA is composed of a resistor R_2 and two capacitors c_1 and c_2, which forms two poles and one zero, and its frequency domain characteristics can be expressed as:

$$H(jw)=\frac{1+j\omega c_2\times R_2}{j\omega(c_1+c_2)+(j\omega)^2 c_1\times c_2\times R_2} \quad (3)$$

It can be seen that the corresponding zero point is $\frac{1}{c_2\times R_2}$ and the poles are 0 and $-\frac{c_1+c_2}{c_1\times c_2\times R_2}$. The zero can be inserted to eliminate the poor phase margin, which caused by the second pole, to increase the stability of the circuit.

IMPROVED TOPOLOGY

The common non-isolated BUCK topology is shown in Figure 5. V_{IN}, as the input voltage, is applied on drain terminal of power transistor. Based on the detection voltage of feedback resistor R, the power transistor is used to adjust conduction time.

In addition, the power transistor in Figure 5 is located at high-level, and it adopts gate drive technology of the bootstrap circuit, ignoring the loss of power transistor. When V_{IN} is relatively high, gate voltage should be at least $V_{IN} + V_{TH}$. The transistor has high risk of breakdown. Compared with the topology shown in Figure 5, an improvement has been made in Figure 6, in which the power transistor is placed at low level. In this way, the low level N-MOSFET is much safer than high level ones.

As shown in Figure 5, in the traditional Buck topology, considering some key problem as the heat dissipation, maintenance or the module replacement, most components in controller are separated from HB LED. The current sense resistor R will be placed near the ground, and it only needs single-end detection circuit instead of complex differential detection circuit. So it can prevent electromagnetic interference.

In the system shown in Figure 6, the sensor resistor R is placed between source terminal and ground. Its power efficiency is increased compared to the circuit shown in Figure 5 [6] . In addition, in Figure 6, the lowlevel N-MOSFET and sense resistor R can transmit inductive current only during part of the cycle, but the power loss of R in Figure 5 occurs during the entire cycle, so the power loss of R in Figure 6 equals to that of R in

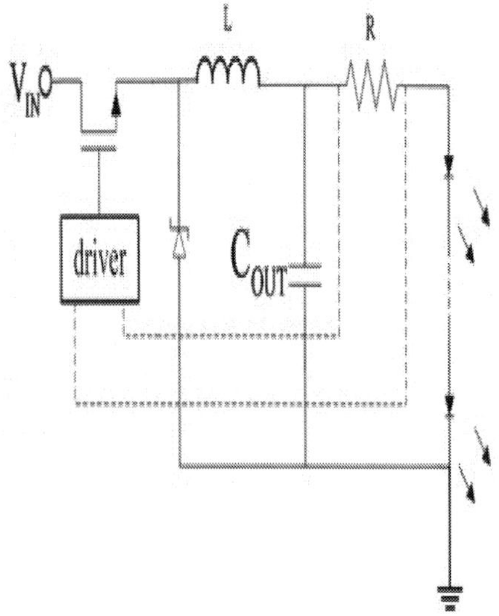

Figure 5. Traditional BUCK topology.

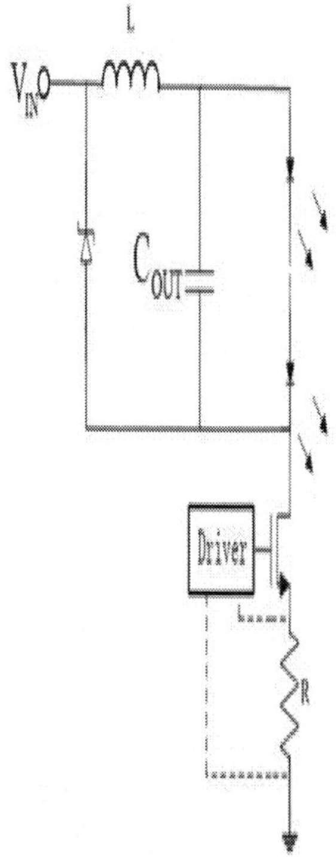

Figure 6. Optimized structure of LED driver.

Figure 5 multiplied by the switch duty cycle D, and the value of D is usually lower than one. Therefore, in Figure 6, the power loss of R will be reduced during a switching period $(1-D)T$, which can be shown as:

$$P = \frac{1}{T}\int_0^{DT} i_{\mathrm{LED}}^2(t)R\mathrm{d}t \tag{4}$$

In Figure 5, the power loss of R can be expressed as:

$$P' = \frac{1}{T}\int_0^{T} i_{\mathrm{LED}}^2(t)R\mathrm{d}t \tag{5}$$

The saved power can be expressed as:

$$\Delta P = \frac{1}{T}\int_0^{(1-D)T} i_{\mathrm{LED}}^2(t)R\mathrm{d}t \tag{6}$$

According to the characteristic of the inductor, the above formula can be further shown as:

$$\Delta P = \frac{1}{T}\int_0^{(1-D)T}\left(I_{\text{peak}} - \frac{V_L}{L}t\right)^2 R\mathrm{d}t$$
$$= R(1-D)\left[I_{\text{peak}}^2 - \frac{V_L I_{\text{peak}}}{L}T(1-D) + \frac{1}{3}\frac{V_L^2}{L}T^2(1-D)^2\right] \quad (7)$$

where T is the switching cycle, R is the sense resistor, and I_{peak} indicates the peak current of the inductor L, and V_L is the voltage across the inductor L. When the transistor is off-state, during $(1-D)T$, V_L equals the voltage drop across the LED lights. It can be seen from the above equation that placing the sense resistor on the low side will reduce the power loss.

The schematic shown in Figure 7 is a complete LED driver circuit structure. It uses the average current control mode as shown in Figure 3, which increases the accuracy of the output current. At the same time, the optimized topology as shown in Figure 6 will reduce the overall power loss.

SIMULATION RESULT AND LAYOUT DESIGN

This design is implemented in CSMC 0.5 µm 40 V BCD technology. Figure 8 is the simulation result of gain and phase margin of error amplifier. By the reasonable settings of c_1, c_2, R_2 in Figure 3 to adjust the location of zero, the stability of the circuit is enhanced. From the diagram it can be seen that the gain of the error amplifier is 60 dB, and the phase margin, which ranges from 180 to 47.75 degrees, is about 132 degree.

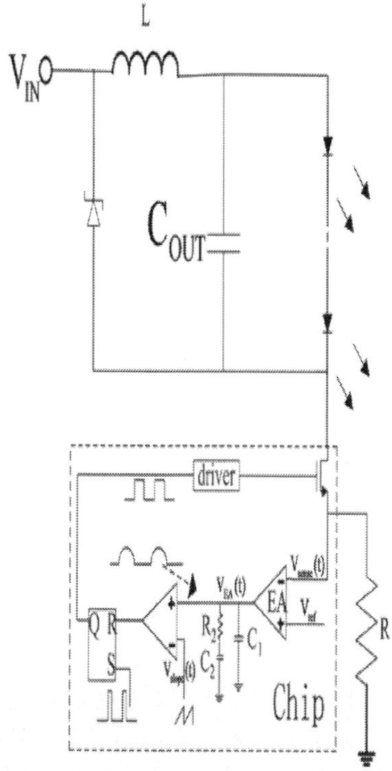

Figure 7. Complete LED driver circuit structure, the part shown within dashed line denotes the designed integrated circuit.

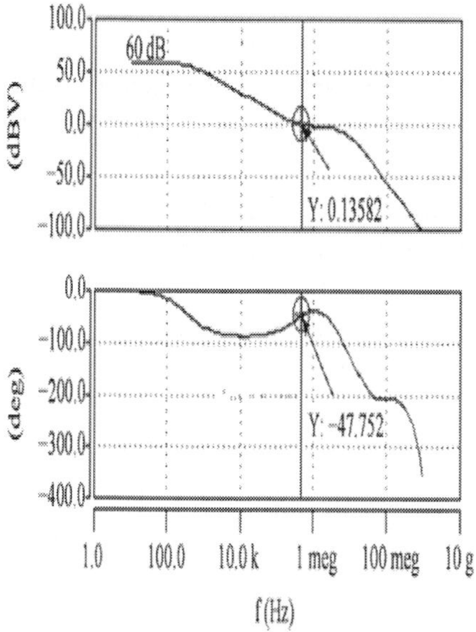

Figure 8. Bode plot of the core circuit.

Figure 9 shows the driving current curve through the LED lamp. Based on the BUCK topology inFigure 6, a suitable inductor L should make the ripple current amplitude as small as possible. The greater inductor will produce more accurate output current. However, in order to minimize the physical size of the circuit, the value of inductor should be selected to enable the circuit working in continuous conduction mode. At the same time, it should be guaranteed that the peak current of the inductor does not exceed its saturation current. Calculation result shows that the value of L is over 22 μH. As shown in Figure 9, in which L equals to be 22 μH, C_{out} equals 10 μF, V_{in} equals 32 V, and eight 1 W of HB LED lights acting as load. It can be seen from Figure 9 that the ripple of transient output current is 4.8 mA and the error is about 0.68%, indicating that the output current is stable and has high accuracy.

Figure 10 corresponds to the situation as same as that in Figure 9 except that L equals 44 μH. The ripple of the transient output current is 1.6 mA, the error is about 0.23% with unchanged average current.

Based on the values shown in Figure 9 and Figure 10, although the difference of the inductance value is twice, its average current remains constant, and the ripples are small. For the traditional peak current mode, its average current and ripple depend heavily on the value of inductor L. So, the average current mode has good stability to the output current, not changing with the value of the inductor L.

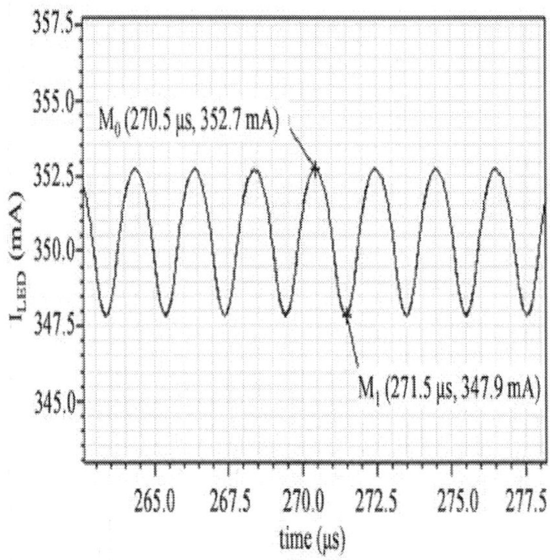

Figure 9. Output drive current ripple when L = 22 μH.

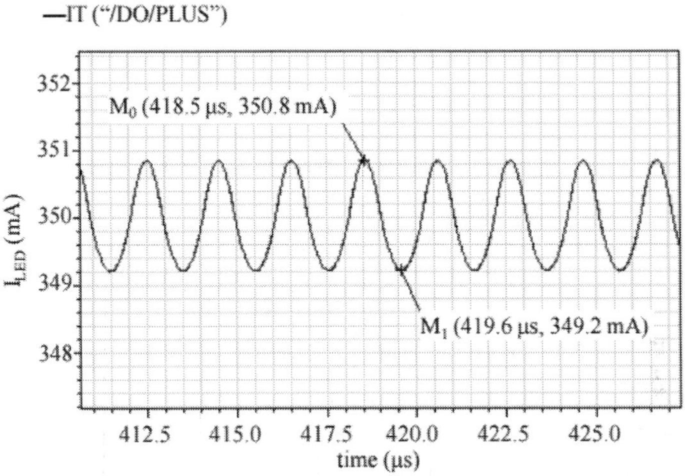

Figure 10. Output drive current ripple when L = 44 μH.

Figure 11 shows the efficiency of HB LED driver circuit. Taking into account the heat dissipation problem of HB LED lights, the simulation curve displays the efficiency when the ambient temperature is 90 degrees celsius. The abscissa denotes the input voltage V_{in}, and the vertical axis is the efficiency. Three curves corresponds different loads from top to bottom, which are eight LED lights, six LED lights and two LED lights. It can be seen that all efficiency are above 91% and the highest is above 96%.

Figure 12 shows the layout of this design. Because the power transistor is integrated into the chip, its power consumption is converted into heat during normal working. During the chip layout, the power MOS transistor should be placed at the edge of chip and an isolation belt has been placed to decrease interference to other circuits. The power supplies of digital and analog circuits should be separated. Considering its larger transient voltage and current in digital circuit, in order to reduce its interference in analog circuits, it is necessary to separate the power supply. It is also paid attention to the total match of difference pair, and the match between resistors and capacitors and so on. The final layout size is 1.4 mm × 1.4 mm.

Using the LED driver chips used in the external topology, the system is verified, mainly focusing on the accuracy of output current and the efficiency at different L values, varied from 22 μH to 55 μH. The number of LED lights remains eight all the time while V_{in} remains 32 V and C_{out} is fixed to be 10 μF. The measured result is shown in Figure 13. The output currents remain almost same under different inductors, totally within 0.3% error.

Table 1 shows the test results of the average current and efficiency, in which L = 22 μH, C_{out} = 10 μF, V_{in} = 32 V, and the numbers of 1 W LED lights are 2, 6, 8, respectively.

The average currents, corresponding to different number of LED, are larger than expecation value. The error is less than 1%, which may be caused by the process variation and thermal factors, etc.

The efficiency is 96% with 8 LED lights, 95% with 6 lights and 92% with 2 lights, respectively. The data shows the high efficiency of the designed integrated circuit and the application topology.

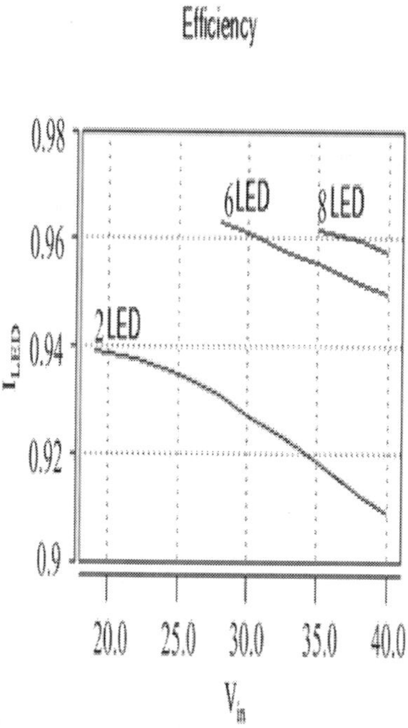

Figure 11. The efficiency of HB LED driver circuit.

Figure 12. Layout design of the chip.

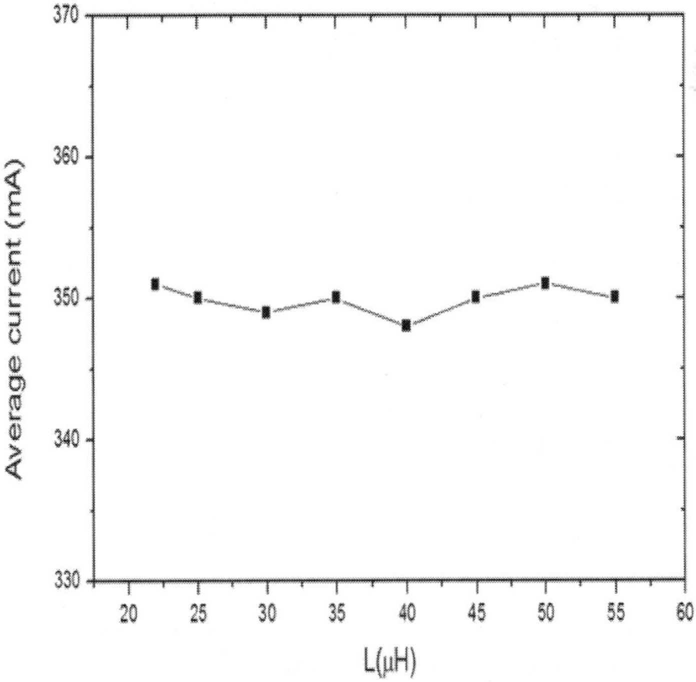

Figure 13. Relationship between average current and the inductance L.

Table 1. Main parameters of the test results.

Test parameters	Num. of LED lights	Size	Unit
Average current	2	352	mA
	6	351.5	
	8	350.6	
Efficiency	2	92	%
	6	95	
	8	96	

CONCLUSION

A LED lighting driver has been designed, including an integrated circuit with integral part and the buck converter topology with low-level switch and low-side sense resistor. The integrated circuit has been fabricated based on the CSMC 0.5 μm 40 V BCD process. In this way, the driver with high accurate output current and high efficiency can be obtained. The final result shows that the output current error is less than 1% and the efficiency is above 96%. Since a significant improvement has been obtained, the circuit is potentially applicable for future LED lighting.

ACKNOWLEDGEMENTS

This project is supported by Beijing Natural Science Foundation (4122031) and Natural Science Foundation of China (60876078).

REFERENCES

1. Leung, W.Y., Man, T.Y. and Chan, M. (2008) A High-Power-LED Driver with Power-Efficient LED Current Sensing Circuit. Proceedingsof Solid-State Circuits Conference, Edinburgh, 15-19 September 2008, 354-357.
2. Ma, F.F., Chen, W.Z. and Wu, J.C. (2007) A Monolithic Current-Mode Buck Converter with Advanced Control and Protection Circuits. IEEE Transactions on Power Electronics, 22, 1836-1846. http://dx.doi.org/10.1109/TPEL.2007.904237
3. van Heinz, B., Georg, S. and Matthias, W. (2007) Power Driver Topologies and Control Schemes for LEDs. APEC 2007—Twenty Second Annual IEEE, Anaheim, 25 February-1 March 2007, 1319-1325.

4. Deng, H., Duan, X., Sun, N., Ma, Y., Huang, A.Q. and Chen, D. (2005) Monolithically Integrated Boost Converter Based on 0.5-μm CMOS Process. IEEE Transactions on Power Electronics, 20, 628-638.http://dx.doi.org/10.1109/TPEL.2005.846551
5. Lee, C.F. and Mok, P.K.T. (2004) A Monolithic Current-Mode CMOS DC-DC Converter with On-Chip Current-Sensing Technique. IEEE Journal of Solid-State Circuits, 39, 3-14.http://dx.doi.org/10.1109/JSSC.2003.820870
6. Benini, L., Bogliolo, A. and De Micheli, G. (2000) A Survey of Design Techniques for System-Level Dynamic Power Management. IEEE Transactions on Very Large Scale Integration Systems, 8, 299-316.http://dx.doi.org/10.1109/92.845896

CITATION

Jiang, Y. , Zhang, D. and Yu, M. (2014) A Novel Integrated Circuit Driver for LED Lighting. *Circuits and Systems*,**5**, 161-169. doi: 10.4236/cs.2014.57018.

Chapter 10

A Vibration-Based MEMS Piezoelectric Energy Harvester and Power Conditioning Circuit

Hua Yu [1,2,3,*], *Jielin Zhou* [1,2,3], *Licheng Deng* [1,2,3] *and Zhiyu Wen* [1,2,3]

[1]College of Optoelectronic Engineering, Chongqing University, Chongqing 400044, China
[2]Key Laboratory for Optoelectronic Technology & Systems, Ministry of Education of China, Chongqing 400044, China
[3]National Key Laboratory of Fundamental Science of Micro/Nano-Device and System Technology, Chongqing 400044, China

ABSTRACT

This paper presents a micro-electro-mechanical system (MEMS) piezoelectric power generator array for vibration energy harvesting. A complete design flow of the vibration-based energy harvester using the finite element method (FEM) is proposed. The modal analysis is selected to calculate the resonant frequency of the harvester, and harmonic analysis is performed to investigate the influence of the geometric parameters on the output voltage. Based on simulation results, a MEMS $Pb(Zr,Ti)O_3$ (PZT) cantilever array with an integrated large *Si* proof mass is designed and fabricated to improve output voltage and power. Test results show that the fabricated generator, with five cantilever beams (with unit dimensions of about 3 × 2.4 × 0.05 mm^3) and an individual integrated Si mass dimension of about 8 × 12.4 × 0.5 mm^3, produces a output power of 66.75 μW, or a power density of 5.19

$\mu W \cdot mm^{-3} \cdot g^{-2}$ with an optimal resistive load of 220 kΩ from 5 m/s^2 vibration acceleration at its resonant frequency of 234.5 Hz. In view of high internal impedance characteristic of the PZT generator, an efficient autonomous power conditioning circuit, with the function of impedance matching, energy storage and voltage regulation, is then presented, finding that the efficiency of the energy storage is greatly improved and up to 64.95%. The proposed self-supplied energy generator with power conditioning circuit could provide a very promising complete power supply solution for wireless sensor node loads.

INTRODUCTION

The use of wireless sensors and implanted medical electronic systems has developed rapidly in recent years. They are usually powered by standard batteries which become depleted within a relatively short timeframe, but the replacement or recharge of batteries is a major bottleneck for wide deployment of wireless sensor nodes (WSN). Moreover, while the size of electronic circuitry has shrunk thanks to the advent of integrated circuit technology, batteries are nowadays often the most bulky devices in wireless sensor nodes. Vibration-based MEMS power generators present an amazing solution to supply power for wireless sensor nodes, which can generate mW or μW level power. Vibration energy harvesting has been exploited for converting ambient kinetic energy into electric energy by several different transduction mechanisms, including piezoelectric, electromagnetic and electrostatic methods. Among these three energy harvesters, the piezoelectric energy harvester (PEH) has a high electromechanical coupling effect, and requires no external voltage sources, it is compatible with MEMS technology, and accordingly has received much recent attention [1–4].

The piezoelectric material chosen for this research is PZT, because it is a well-known and well-characterized material with a high piezoelectric coefficient. Vibration energy harvesters must be able to respond to the low frequency and low acceleration vibrations that usually exist in the environment. Moreover, the energy harvester should generate as much energy as possible in order to supply enough power for the follow-up loads. Recently, MEMS technology has been applied toward the development of energy harvesters, and many piezoelectric MEMS energy harvesters have been developed. Renaud *et al.*proposed a fabricated MEMS-based PZT cantilever micro-generator with an integrated proof mass that can generate an average power of 40 μW at 1.8 kHz vibration frequency [5]. Jeon *et al.* developed a *d33*mode thin film PZT power generating device with interdigitated electrodes that can generate an average power of 1.0 μW from 108 m/s^2 vibration acceleration at its resonant frequency of 13.9 kHz [6]. However, the above two energy harvesters' resonant frequencies are very high. Fang *et al.* fabricated a MEMS-based PZT cantilever power generator with a nonintegrated Ni proof mass that can generate 2.16 μW from 10 m/s^2 vibration acceleration at its resonant frequency of 609 Hz. The nickel metal mass on the tip of the cantilever is used to decrease the structure's resonant frequency for the application under low-frequency vibration, but it cannot to be micro-machined by MEMS technology [7]. Similar to the previous work, Liu *et al.* used the previous cantilever structure to construct a power generator array to improve power output and frequency flexibility. Although they demonstrated that power density was high, the proof mass was not also integrated with the cantilever which will be an additional difficulty in production [8]. Muralt *et al.* designed and fabricated a micro power generator of thin film PZT laminated cantilever with proof mass and interdigitated electrodes which could generate a voltage of 1.6 V and power of 1.4 μW when excited under 20 m/s^2 vibration acceleration at 870 Hz resonant frequency [9]. Over the past two years, some new structures or new piezoelectric materials have been applied in energy harvesters, which attain lower resonant frequency and more output power, but the power conditioning circuit for the improvement of high internal output impedance about piezoelectric energy harvester has rarely been discussed in these references [10–12]. From these previous studies, at least three conclusions can be reached and summarized as follows: Firstly, most MEMS vibration energy harvesters usually work under high frequency environment conditions, which limits the scope of application for these devices, because the frequencies of available ambient vibration sources are relatively low. Secondly, the delivered power output of those MEMS micro-generators is relatively low, so the architecture and parameters of energy harvester must be optimized as far as possible in order to increase output power. Thirdly, a power conditioning circuit is absolutely necessary for piezoelectric energy harvesters in order to extract the maximum power from the harvester, store energy and then regulate the voltage for meeting the demands of the loads, such as wireless sensor nodes [13].

In this paper, a complete design flow of a self-powered MEMS PZT vibration energy harvesting system for wireless sensor nodes is presented, which includes the design method of how to increase the output power and voltage of the energy harvester. A design method of low resonant frequency for the vibration energy harvester is also verified. This paper especially gives the theoretical calculation formula of the resonant frequency, the simulation value using ANSYS FEM software, and the experimental test results. It's shown that their relative errors are very small, which provided a reference design method of the resonant frequency for the vibrational energy harvester. Moreover, the total output power and voltage are greatly improved by improving the PZT film material preparation technology and MEMS cantilever array fabrication process, which lays a strong foundation of the latter power condition circuit. Finally, in order to improve the high internal impedance characteristics of the PZT generator, we design an efficient autonomous power conditioning circuit, with the functions of impedance matching, energy storage and voltage regulation, finding that the efficiency of the energy storage is greatly improved. The proposed whole energy harvesting system could provide a complete self-powered supply solution to wireless sensor node loads, which is regarded as a reference method for implementing a battery-less wireless sensor network.

DESIGN OF THE MEMS PIEZOELECTRIC ENERGY HARVESTER

Architecture Design

The cross-section structure of the MEMS piezoelectric cantilever array is shown in Figure 1. The cantilever array device is designed to resonate at a specific low vibration frequency by using a large integrated Si proof mass. The bending results in a strain distributed along the beam, which is then converted to electrical energy through the transverse mode (*d31*) piezoelectric effect. The Pt/Ti and Al electrodes are patterned in order to generate strain in parallel to the electric field, which forms the *d31*mode of the piezoelectric element. When an input vibration applies acceleration to the beam structure, the effective mass converts the input acceleration into force. The relative displacement causes the PZT layer to be tensed or compressed, which in turn induces a charge shift and accumulation due to the piezoelectric effect. Electrodes collect the generated charge and electrical damping results. The magnitude of the electric charge voltage is proportional to the stress induced by the relative displacement. Material properties and structural parameters of the piezoelectric energy harvester are listed in Table 1.

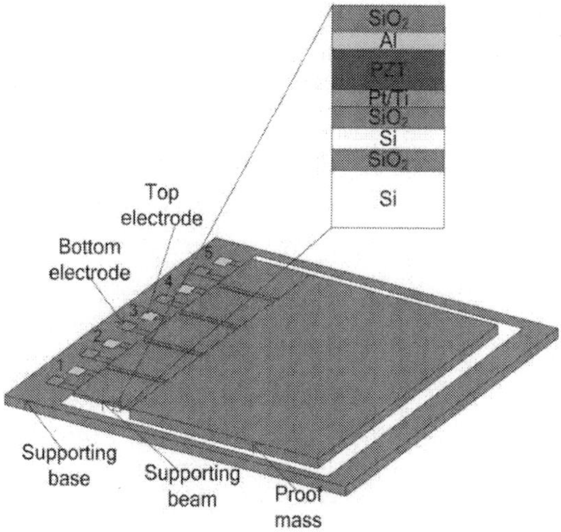

Figure 1. Cross section structure of MEMS piezoelectric cantilever array.

Table 1. Material properties and structural parameters of the piezoelectric energy harvester.

Symbol	Parameter	Value
ρ_S	Density of Si beam	2,329 kg/m^3
ρ_P	Density of PZT	7,500 kg/m^3
E_S	Young's modulus of Si beam	170 GPa
E_P	Young's modulus of PZT	98.4 GPa
ν_S	Poisson's ratio of Si beam	0.28
ν_P	Poisson's ratio of PZT	0.29
d_{31}	Piezoelectric constant of PZT	-274×10^{-12} C/N
ε_0	Vacuum dielectric constant	8.85×10^{-12} F/M
$\varepsilon^s_{11}/\varepsilon_0$	Relative dielectric constant	1,703
$\varepsilon^s_{33}/\varepsilon_0$	Relative dielectric constant	1,433.6
$l_S \times w_S \times h_S$	Dimensions of Si beam	$3 \times 2.4 \times 0.05$ mm^3
$l_P \times w_P \times h_P$	Dimensions of PZT	$3 \times 2.4 \times 0.002$ mm^3
w_g	Gaps between the beams	100 μm
$l_m \times w_m \times h_m$	Dimensions of the proof mass	$8 \times 12.4 \times 0.5$ mm^3

2.2. Output Power

Mechanical power is converted into electrical power when damping is present. For a sinusoidal excitation vibration, the electrical power generated by the system is given in Equation (1) [14]:

$$P = \frac{m\zeta Y^2 \left(\frac{\omega}{\omega_n}\right)\omega^3}{\left[1-\left(\frac{\omega}{\omega_n}\right)^2\right]^2 + \left[2\zeta\frac{\omega}{\omega_n}\right]^2} \tag{1}$$

where ζ is the damping ration ($\zeta = q/2m\omega_n$); ω_n and Y is the resonant frequency and the amplitude of vibration respectively, and ω is the excitation frequency. It's obvious that the maximum output power occurs at the resonant frequency of the generator if ω is equal to ω_n for the system without damping:

$$P_{max} = P(\omega_n) = \frac{mY^2{\omega_n}^3}{4\zeta} \tag{2}$$

It also can be concluded from Equation (2) that the value of the maximum power is indeed finite, and reduction of the damping factor results in increased mass displacement, which is ultimately limited by the size and geometry of the device. Equation (3) also expresses the power in terms of the excitation acceleration magnitude of the input vibration:

$$P_{max} = \frac{mA^2}{4\zeta\omega_n} \tag{3}$$

where A is excitation acceleration magnitude, and its formula is A=ω2nY. Several conclusions may be drawn from the above equations. Firstly, if the resonant frequency of the generator is set equal to the proof mass frequency, the maximum power generated is limited by the movement of the proof mass. Thus, the maximum power can be expressed in terms of maximum distance the proof mass can move. Secondly, for applications where the frequencies of vibration are well defined and concentrated around one point, a low damping factor will provide higher peak responses and power generation. Thirdly, it can also be seen that power is proportional to the proof mass, so a large proof mass is always desirable for energy harvesting. Finally, the harvester's damping ζ is composed of the mechanical damping ζ_m and electrical damping ζ_e. The maximum electrical output power is equal to half of the value in Equation (2)when the electrical damping matches the mechanical damping [15].

Resonant Frequency

The resonant frequency of a vibration energy harvester is one of the most important design parameters, because the maximum output power can be obtained when the vibration frequency matches the resonant frequency. It is well-known that the power output will be dramatically reduced when the driving vibration frequency deviates from the resonant frequency of a device [16]. When a horizontal cantilever beam is subjected to a vertical harmonic excitation acting at the tip of the beam, the equivalent stiffness is obtained from the following Equation (4):

$$k_{eq} = \frac{3EI}{L^3} \tag{4}$$

where E is the equivalent elastic modulus, I is the equivalent rational inertia, and L is the effective length of the cantilever beam. The value of EI can be calculated by Equation (5) [17]:

$$\mathrm{EI} = E_{p1} W_p h_p \left[(z_{pN} - z_N)^2 + \frac{h_p^2}{12} \right] + E_{s1} W_s h_s \left[(z_{sN} - z_N)^2 + \frac{h_s^2}{12} \right] \tag{5}$$

where z_{pN} and z_{sN} are the neutral axis of the piezoelectric layer and silicon substrate, z_N is the neutral axis of the total piezoelectric cantilever. They are calculated by Equations (6)–(8) respectively:

$$z_{pN} = \frac{1}{2} h_p \tag{6}$$

$$z_{sN} = h_p + \frac{1}{2}h_s \tag{7}$$

$$z_N = \frac{E_{p1}h_p z_{pN} + E_{s1}h_s z_{sN}}{E_{p1}h_p + E_{s1}h_s} \tag{8}$$

where E_{p1} and E_{s1} are the Young's modulus of PZT material and Si beam under the plane strain respectively, they are expressed by Equations (9) and (10) as follows:

$$E_{p1} = \frac{E_p}{1-\nu_p^2} \tag{9}$$

$$E_{s1} = \frac{E_s}{1-\nu_s^2} \tag{10}$$

where E_p and E_s are the elastic modulus of PZT material and Si material respectively, v_p and v_s are the Poisson's ratio of the PZT layer and silicon cantilever beam layer. Based on the equivalent model of the cantilever beam, the resonant frequency of the piezoelectric cantilever beam can be written as a simpleEquation (11) [16]:

$$f_n = \frac{\omega_n}{2\pi} = \frac{1}{2\pi}\sqrt{\frac{3EI}{(m+\frac{33}{140}m_s)L^3}} \tag{11}$$

where L is the equivalent length of the beam and the mass, m is the mass of the proof mass, and m_s is the mass of the piezoelectric cantilever.

FINITE ELEMENT METHOD MODEL

The proposed piezoelectric energy harvester is analyzed using FEM to predict the relationships between output characteristics and geometries of the harvester. The piezoelectric energy harvester is modeled by the

ANSYS14.0 software. Element types of the elastic base and piezoelectric patches are selected as 'SOLID45' and 'SOLID5', respectively. A mapped meshing method is specified to mesh the created model. In addition, the boundary conditions are applied to the clamped end of the cantilever by applying zero displacement for all degrees of freedom at the nodes. The top and bottom electrodes of the five piezoelectric patches are defined respectively [18]. Figure 2 shows the created finite element model of the piezoelectric energy harvester.

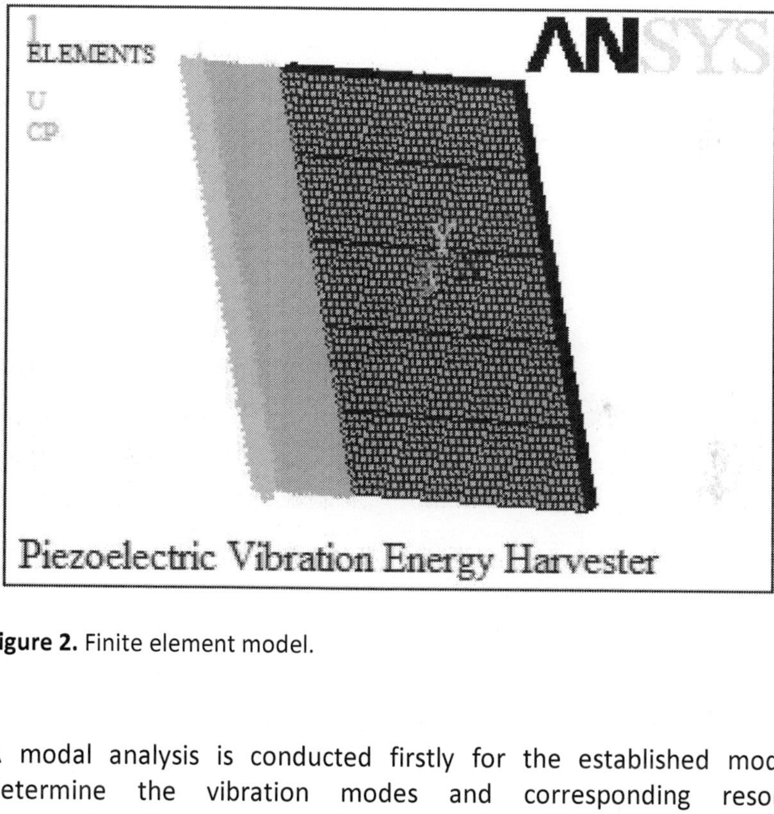

Figure 2. Finite element model.

A modal analysis is conducted firstly for the established model to determine the vibration modes and corresponding resonance frequencies. The resonance frequencies for the first four models are 222.3 Hz, 2,138.6 Hz, 4,057.4 Hz and 28,886.2 Hz, respectively, simulated by using ANSYS software as shown inFigure 3. The horizontal color bar refers to the modal displacements of each point in the structure, and its unit is meters. However, the modal displacements here are not really physical displacements but the relative displacements, and they illustrate the vibration characteristics of the structure. The first resonant frequency (222.3 Hz) of the model makes it possible to use it in the low frequency range, which is always chosen for dynamic analysis to achieve the maximum power. Besides, there is a quite difference between the first frequency and the others, which ensures the stability of the piezoelectric energy harvester.

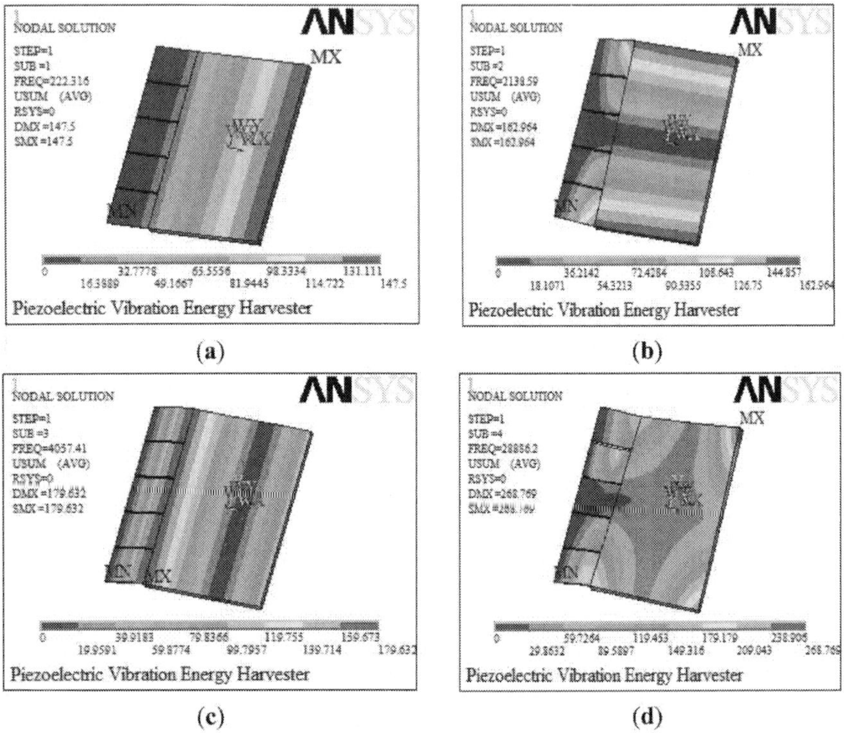

Figure 3. The first four resonant frequencies of the piezoelectric energy harvester. (**a**) The first resonant frequency. (**b**) The second frequency. (**c**) The third resonant frequency. (**d**) The fourth resonant frequency.

Secondly, a harmonic analysis is implemented to calculate the response of the structure to cyclic loads over a frequency range, which can provide the curves between the induced voltage and the dynamic frequency. This can help the designer evaluate the dynamic traits of the results and any possible fatigue and crack problems to maintain the stability of the device. In the harmonic response analysis process, sinusoidal signal $z(t) = u_0 sin(\omega t)$ is forced on the whole structure in the poling direction of the piezoelectric patch. Figure 4 shows the curves between the induced voltage and the dynamic frequency in four cases. Case 1 and Case 2 show the electric potential of single beam and five series connected beams at 5 m/s^2 vibration acceleration, respectively. Case 3 and Case 4 present separately the output voltages derived from single beam and five connected beams in series at 10 m/s^2 vibration acceleration. All four simulation results indicate that maximum electric potential occurred at the first resonance frequency of 222.3 Hz and then declined rapidly. That is to say, the piezoelectric cantilever is only sensitive to the harmonic frequency. If the acceleration is twice as much as before, the value of output voltage is doubled, which is mainly due to the fact that harmonic analysis is a linear analysis. Besides, the output voltage of five beams connected in series happened to be nearly five times as much as that of single beam as a result of the symmetry of the piezoelectric vibration structure. The output voltage response of piezoelectric energy harvester can be seen distinctly under differing frequency exciting forces by the harmonic response, which provides the theory for further systematic dynamic analysis and the structure design of the energy harvester. These also provide an applied dynamic force analysis method, which can be used as reference for other analysis.

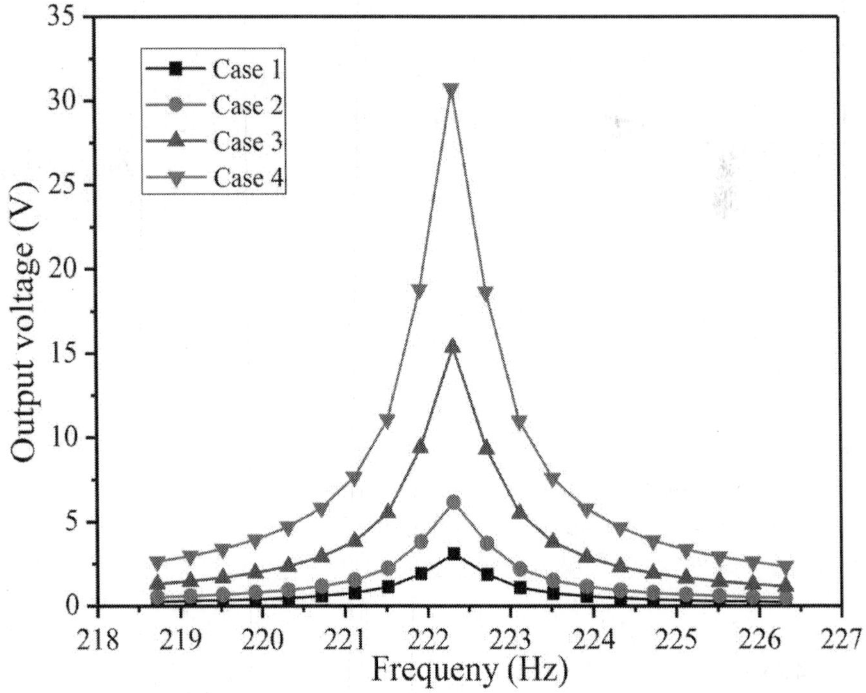

Figure 4. Harmonic curves.

Thirdly, the paper also studies how the resonant frequencies of the piezoelectric energy harvester structure are affected by the length, the width and the thickness of the cantilever beam. Different simulation results are plotted in Figure 5. Figure 5a shows that when the width and thickness of the cantilever beam are fixed, the resonance frequency decreases with the increase of the length of the cantilever beam. As shown in Figure 5b, with the increasing width of the cantilever beam, there is a clear growth trend in the resonance frequency.

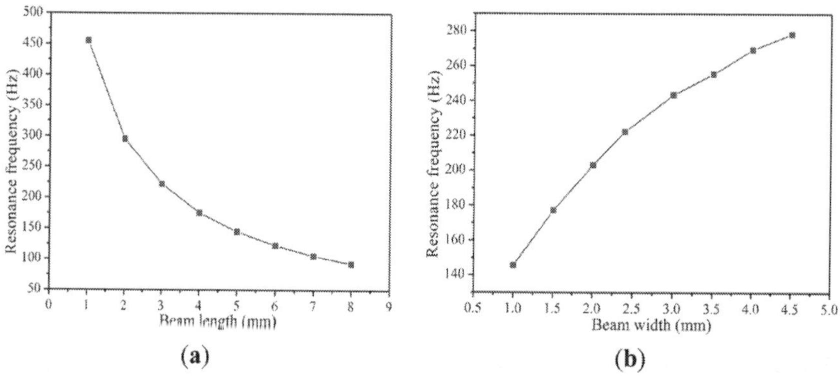

Figure 5. Resonance frequency versus dimensions of the beam. (**a**) Beam length. (**b**) Beam width. (**c**) Beam thickness.

In Figure 5c, it can be seen that the resonant frequency increases nearly linearly with the thickness. Accordingly, in order to make the piezoelectric energy harvester work in a low vibration frequency environment, the cantilever beam should be designed to be longer, narrower and thinner on the condition that it cannot be fractured.

Figure 6 presents the relationship between resonance frequency of the piezoelectric energy harvester and the mass at the tip of the beam. The resonance frequency decreases sharply from 508.0 Hz to 165.53 Hz when the mass of the proof mass increases from 0.02 g to 0.2 g. As a result, by adjusting the mass of the proof mass, the resonance frequency of the piezoelectric energy harvester can be significantly changed to make it work better to meet the low frequency vibration needs.

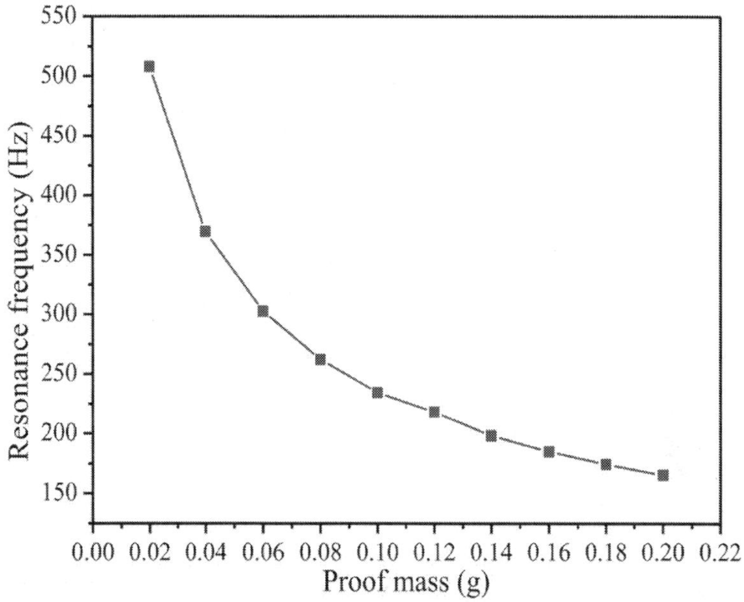

Figure 6. The resonance frequency *versus* the proof mass.

It is obvious that the targeted resonant frequency is attained by changing the lengths and widths of the cantilever beam and the dimensions of the proof mass. Based on the simulation results, the structural parameters of the piezoelectric energy harvester in Table 1 are chosen in this design. Substituting the involved parameters into Equation (5), the value of EI is as follows: $EI = 4.95 \times 10^{-6}$. Then, substituting the value of EI, L, m, m_b into Equation (11), the theoretical value of resonance frequency is as shown inEquation (12). Compared the theoretical value of resonance frequency with the simulation resonance frequency of Figure 4, the relative error value between theoretical and simulation resonant frequency is only 2%. This result shows the theoretical analysis has high prediction accuracy, which provides the basis of architecture parameters design for MEMS PZT energy harvester:

$$f_n = \frac{\omega_n}{2\pi} = \frac{1}{2\pi}\sqrt{\frac{3EI}{(m + \frac{33}{140}m_s)L^3}} = 217.65\,(Hz) \tag{12}$$

Finally, to investigate the dependence of the output voltage on the parameters of the piezoelectric patch, additional harmonic analyses are done. Three plots of the relationships between them are shown inFigure 7. Figure 7a shows the dependence of the output voltage on the length of the piezoelectric patch. While other parameters remain constant, the output voltage first increases and then decreases with the increase of the length of piezoelectric patch. From Figure 7b, it is found that the output voltage decreases when the width of the piezoelectric patch is changed from 0.25 mm to 2.4 mm. The result in Figure 7cshows that the output voltage begins to increase first, and then decreases with the growth of the thickness of the piezoelectric patch. The analysis results show that the piezoelectric patch with the appropriate length and thickness should be less wide compared with the dimensions of substrate, which is well in agreement with the previous literatures [19].

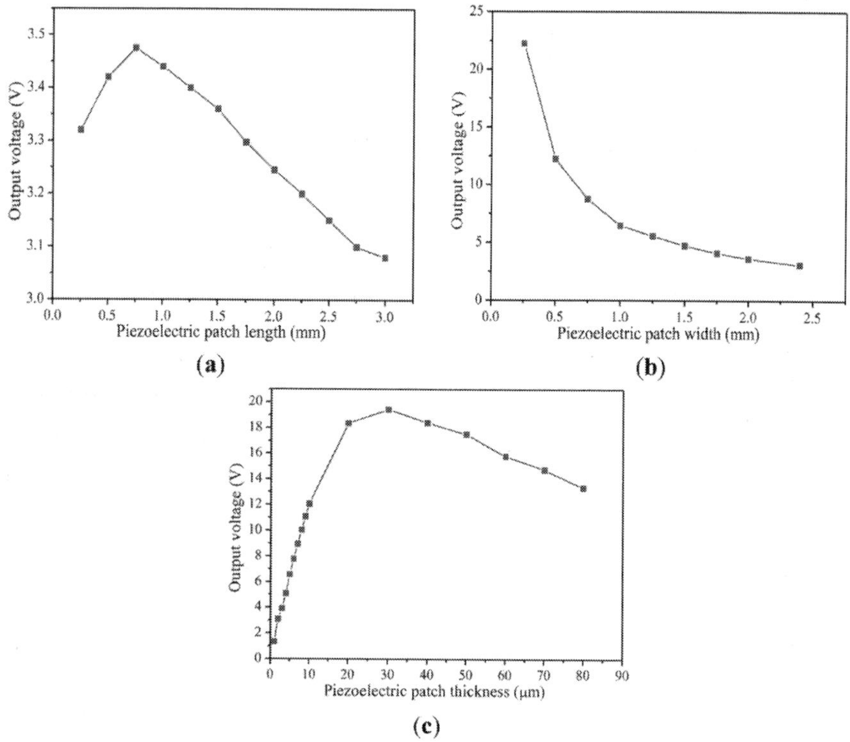

Figure 7. Output voltage versus piezoelectric patch dimensions. (**a**) Patch length. (**b**) Patch width. (**c**) Patch thickness.

Based on these simulation results, a MEMS piezoelectric cantilevers array (five beams of 3 × 2.4 × 0.05 mm^3) integrated with a large silicon proof mass of about 8 × 12.4 × 0.5 mm^3 is designed to improve the output voltage for the low frequency vibration condition.

FABRICATION

The MEMS piezoelectric energy harvester consists of five piezoelectric PZT cantilevers and an integrated silicon proof mass. The five PZT elements are electrically isolated from one another and each PZT element is composed of a top electrode layer (Al), a PZT layer, and a bottom electrode layer (Pt/Ti). Each of the top and bottom electrodes of the PZT element are connected to a bonding pad individually. A large proof mass is designed and integrated at the end of the supporting beam to achieve a low resonant frequency and increase output power. The fabrication process is as shown in Figure 8 [20,21]:

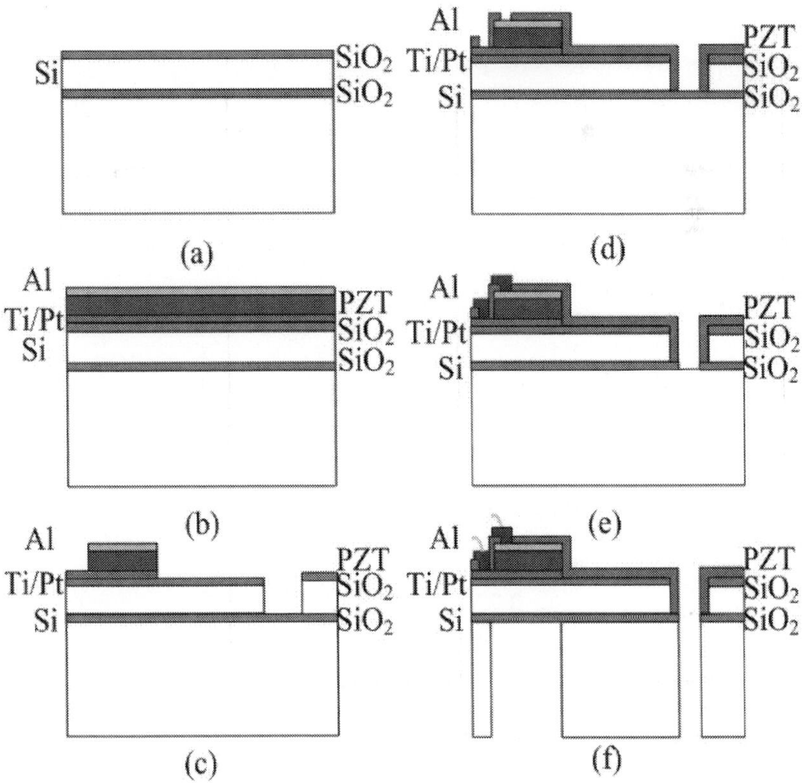

Figure 8. Fabrication process of MEMS piezoelectric energy harvester.

a) Grow thermal oxide on silicon on insulator (SOI) wafer.
b) Sputter deposit Ti/Pt (bottom electrode), and *Al* (top electrode), spin coat sol-gel PZT piezoelectric layer.
c) Pattern and etch electrode, PZT and SiO_2 layer, deep reactive ion etching (DRIE) Si device layer.
d) Plasma enhanced chemical vapor deposition (PECVD) deposit oxide, pattern and etch via.
e) Pattern and etch contact pads.
f) DRIE from backside, dicing and releasing.

POWER CONDITIONING CIRCUIT

The main goal of the power conditioning circuit is highly efficient energy transfer and energy accumulation between the harvester and the electric load, because the typical output power of a MEMS vibration energy harvester is in the μW range. Figure 9 shows the schematic of the proposed power conditioning circuit, which includes AC-DC rectifying, impedance matching, energy storage, instantaneous bleed-off and voltage regulator circuits. The basic design idea is to enable maximum power extraction from the piezoelectric energy harvester by using the impedance matching circuit, store the energy in a super-capacitor and supply power to the load when enough energy is accumulated in super-capacitor. The different parts of the circuit are analyzed with detailed information as follows.

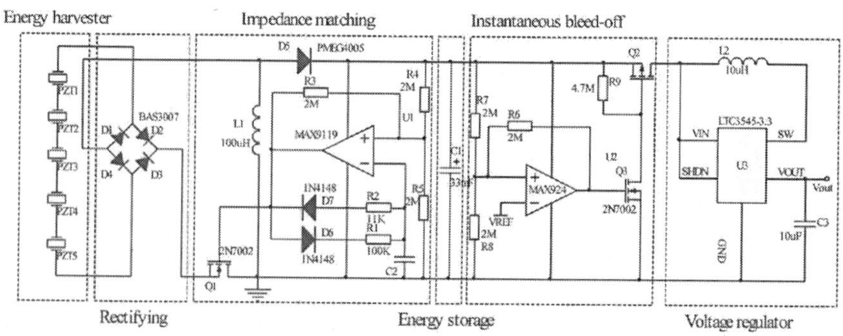

Figure 9. Power conditioning circuit for PZT energy harvester.

Rectifying Circuit

The output of the piezoelectric energy harvester is an alternating voltage signal which must be rectified for supplying power to the sensor node load. The rectifying bridge circuit consists of four small-signal diodes. These diodes are chosen specifically for rectification because they must have the smallest forward voltage drop and leakage current.

Impedance Matching Circuit

The discontinuous control mode (DCM) buck-boost converter is chosen for the second stage because of its ability to accommodate a wide range of input voltage and behave as a lossless resistor to match the source impedance for the maximum power point tracking. The effective input resistance of a DCM buck-boost converter is given in Equation (13) as follows [22]:

$$R_{IN} = \frac{V_{RECT}}{\frac{1}{T_S}\int_0^{D_1T_S} I_{L1}dt} = \frac{2L_1}{D_1^2T_S} \tag{13}$$

where T_s is switching period of the transistor Q_1, D_1 is duty cycle of the transistor Q_1, and V_{RECT} is the rectified voltage. In order to achieve the resistive impedance matching, the effective input resistance R_{IN}should be equal to the optimal resistive load $R_{L,opt}$. Hence, the optimal duty cycle can be expressed by using Equation (14):

$$D_{1,OPT} = \sqrt{\frac{2L_1}{R_{IN,OPT}T_S}} \tag{14}$$

An ultra-low power oscillator with an *RC* network is used to generate the pulse width modulation (PWM) signal for turning the power switch Q_1 on or off. The duty cycle and switching frequency can be adjusted by choosing appropriate R_1, R_2 and C_2 to achieve the maximum power delivery in Figure 9. If the comparator output is high, the voltage at the non-inverting input of the comparator is two-thirds of the supply voltage. Capacitor C_2 is charged through resistor R_1. As soon as the capacitor voltage reaches two-thirds of the supply voltage, the comparator output acts and goes low. The voltage at the non-inverting input of comparator is one-third of the supply voltage. Capacitor C_2 is discharged through resistor R_2. The power supply of the chip connects to the cathode of diode D_5.Once the capacitor voltage decreases one-third of the supply voltage, the comparator output becomes high, Capacitor C_2 is charged again and the whole cycle repeats.

Energy Storage and Instantaneous Bleed-Off Circuit

In order to drive the load successfully, the weak harvested energy should be accumulated in the energy storage element during the long period time because a wireless sensor node requires much more peak power than a piezoelectric micro power generator can produce. A super-capacitor C_1 of 33 mF is chosen as a storage element in this energy harvesting system, which accumulates energy to enable efficient use for short power output bursts. The super-capacitor should be disconnected from the load during the energy accumulation stage to prevent energy leakage to the load, and is supposed to be connected to the load only if the accumulated energy is large enough to drive it. The instantaneous bleed-off circuit is composed of an ultra-low power voltage comparator U_2 with hysteresis, the switching Q_2, and Q_3. The comparator monitors the super-capacitor's voltage and controls the super-capacitor's charging and discharging process. The switch is in charge of powering on or off the wireless sensor node. The hysteresis of a comparator creates two threshold voltage points: the upper one for super-capacitor charge voltage and the other lower one for the super-capacitor discharge voltage, which can control how much energy discharging from the super-capacitor.

Voltage Regulator

A voltage regulator is adopted as the output unit providing a stable voltage supply to the wireless sensor node. In order to improve the efficiency of the voltage regulator, two key design points are applied in the circuit. One is to reduce inductor power loss and the other is set to enable start-up signal for DC-DC converter. If input voltage of the voltage regulator (*i.e.*, the super-capacitor's voltage) is lower than a defined start-up voltage, the voltage regulator consumes unnecessary power because it is not able to boost the output voltage. Thus, to overcome this drawback, we introduce a supervisory unit that continuously checks the super-capacitor's voltage and enables the voltage regulator output stage only when it can be equal to the value of startup voltage. Otherwise, a complete shutdown with output disconnection suppresses any additional power consumption, reducing the charging time of the super-capacitor and boosting the overall efficiency.

TESTING AND ANALYSIS

Experimental Test Setup

In order to measure the output characteristics of the piezoelectric energy harvester, a vibration testing system is employed. A schematic drawing of the experimental setup for PEH system is shown in Figure 10. It consists of a vibration exciter, an accelerometer, a power amplifier, an oscilloscope, a signal generator and a charge amplifier. An accelerometer is assembled on the vibration exciter together with the cantilever array device for the acceleration measurement. The vibration signal is generated from the signal generator, amplified via the power amplifier and finally utilized to control the vibration amplitude and frequency of the shaker. Accordingly, the piezoelectric cantilever device will undergo excitations and generate output voltage signal which is recorded to the oscilloscope. Acceleration signals will be measured by the accelerometer and amplified by the charge amplifier, and then shown on the monitor of the computer. The photo of the experimental system for testing the MEMS piezoelectric energy harvester is shown in Figure 11.

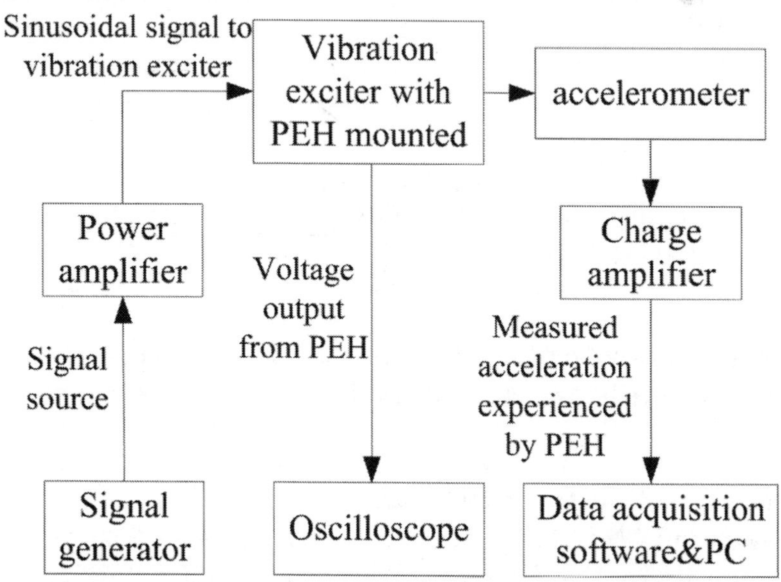

Figure 10. Schematic drawing of the experimental setup for PEH system.

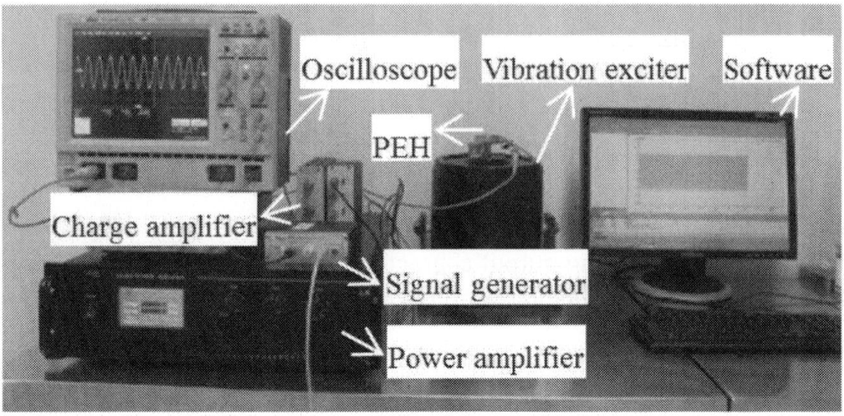

Figure 11. Photo of the experimental setup for MEMS piezoelectric energy harvester.

Test Results and Analysis

First, the respective characteristics of the MEMS piezoelectric energy harvester are measured, including the resonant frequency, output voltage, output power and optimal load. Figure 12a shows the output voltage of a single PZT beam and five beams connected in series as the vibration frequency swept from 231 Hz to 237 Hz at 5 m/s^2 vibration acceleration. As can be seen, the maximum open circuit voltage is 1.62 V and 7.04 V at the resonant frequency of 234.5 Hz under 5 m/s^2 acceleration, respectively. The error between the simulated and measured resonant frequency is about 4.2%. Figure 12bshows the load voltage and average power to the different loads at resonant frequency. As expected, the load voltage increases with the increased load, up to 3.91 V at 230 KΩ load. However, the power delivered to the load has a maximum value. The maximum output power of 66.75 µW is obtained at 5 m/s^2 acceleration with an optimal resistive load of 220 kΩ resulting in a maximum power density of 5.19 $\mu W \cdot mm^{-3} \cdot g^{-2}$. The power density is calculated using the average power divided by the effective volume of the device, which is the volume of the entire beam and the Si proof mass calculated from the measured data in Table 1. Some main performance parameters of recent MEMS piezoelectric energy harvesters are compared in Table 2.

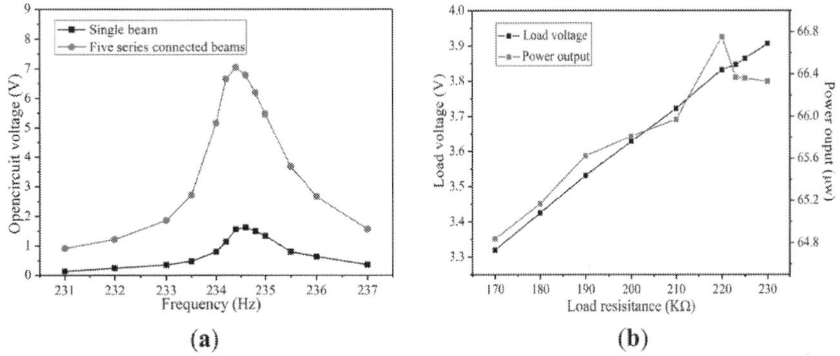

Figure 12. The characteristics of MEMS piezoelectric energy harvester. (**a**) Open circuit voltage *vs*. the frequency. (**b**) Load voltage and power *vs*. the load.

Table 2. Comparison of recent MEMS piezoelectric energy harvesters.

Reference	Material	Volume mm^3	A m/s^2	f Hz	Power µW	Power Density $\mu W \cdot mm^{-3} \cdot g^{-2}$
[23]	PZT d_{31}	1.063	5	243	2.33	8.39
[24]	PZT d_{31}	12.775	9.3	615	51.3	4.97
[25]	PZT d_{31}	27	15	154	205	3.37
[26]	PZT d_{31}	0.608*	10	78.7	0.803	1.32
This paper	PZT d_{31}	51.472	5	234.5	66.75	5.19

*Estimated from reference data.

The next experiment shows the performance of the power conditioning circuit. Figure 12 shows that the proposed power MEMS system can successfully drive the wireless sensor load when the temperature-humidity sensor node transmits signals. As shown in Figure 13, when the voltage across the super-capacitor is charged to 2 V, the instantaneous bleed-off circuit begins to work and the voltage regulator will start up, supplying 3.3 V to the wireless sensor node load. The super-capacitor's voltage will begin to drop while the energy is supplied to the load. When the voltage across the super-capacitor drops below 1.5 V, the output voltage level of instantaneous bleed-off circuit will become low, so the voltage regulator shut off again. When the super-capacitor voltage is charged to 2.0 V next time, the DC-DC converter will start up again. Through the test, the working cycle time of the wireless sensor node is 0.185 h. In the proposed circuit, the energy of super-capacitor discharging is shown in Equation (15) and the efficiency of the energy management is calculated by Equation (16), respectively:

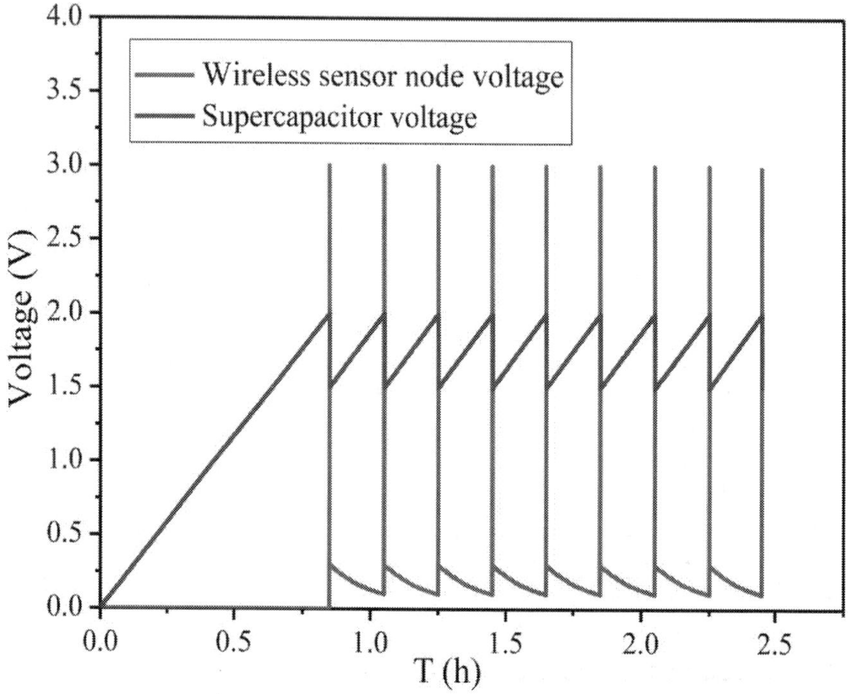

Figure 13. The voltage across the super-capacitor and wireless sensor node.

$$E_{SC} = \frac{1}{2}CV_{THR}^2 - \frac{1}{2}CV_{THF}^2 = \frac{1}{2} \times 33\text{mF} \times 10^{-3} \times (2^2 - 1.5^2)\,\text{V} = 28.875\text{mJ} \quad (15)$$

$$\eta = \frac{P_{stored,SC}}{P_{MAP}} = \frac{E_{SC}/T}{P_{MAP}} = \frac{28.875mJ/0.185h}{66.75\mu W} = 64.95\% \quad (16)$$

The maximum available power of 66.75 μW is measured as the power delivered to the 220 kΩ optimal resistive load connected directly to the piezoelectric generator at the resonant frequency of 234.5 Hz. The extracted power of the proposed power conditioning circuit is the power delivered to the super-capacitor during the charging process. The proposed circuit's efficiency is about 64.95%, which is better than the results reported in some previous literature [27–29].

RESULTS

In this paper, a self-supplied MEMS piezoelectric energy generator with power conditioning circuit is proposed. A complete design flow analyzing the architecture and parameters of the energy harvester using the FEM is established. Based on the simulation results, a MEMS PZT cantilever array with an integrated large Si proof mass is designed and fabricated. Test results show that the proposed energy harvester can produce a maximum output power of 66.75 μW, or power density of 5.19 $\mu W \cdot mm^{-3} \cdot g^{-2}$ with an optimal resistive load of 220 kΩ from 5 m/s^2 acceleration at its resonant frequency of 234.5 Hz, which represents a significant improvement of output power and power density. This paper also offer a design method of a power conditioning circuit with the function of impedance matching, energy storage and voltage regulation in order to improve high impedance characteristics of PZT energy harvester and attain higher efficiency. The experimental results show that the proposed self-supplied energy generator with power conditioning circuit could provide a more promising complete power supply solution for wireless sensor node loads. Future planned work includes further improvement of the energy harvester architecture and development of a monolithic power conditioning chip.

This work is sponsored by the National Natural Science Foundation of China (No. b61074177), the Visiting Scholar Foundation of Key Laboratory for Optoelectronic Technology & Systems Ministry of Education of China, and the Fundamental Research Funds for the Central Universities (No. CDJZR10120006).

AUTHOR CONTRIBUTIONS

The paper was completed in collaboration between all authors. Hua Yu proposed the research theme and idea, performed the theoretical analysis of PEH, completed the circuit design, carried out the laboratory experiments, analyzed the data, interpreted the results and wrote the paper. Jielin Zhou performed the analysis of PEH, conducted the FEM simulations, measured the characteristics of the MEMS PEH, and co-wrote the manuscript. Licheng Deng and Zhiyu Wen designed the fabrication process of the MEMS PEH, co-designed the MEMS PEH testing experiments, and co-worked on associated data collection and their interpretation. All authors have read and approved the final manuscript.

Conflicts of Interest

The authors declare no conflict of interest.

REFERENCES

1. Roundy, S.; Leland, E.S.; Baker, J.; Carleton, E.; Reilly, E.; Lai, E.; Otis, B.; Rabaey, J.M.; Wright, P.K.; Sundararajan, V. Improving power output for vibration-based energy scavengers.*Pervasive Comput.* **2005**, *4*, 28–36.
2. Poulin, G.; Sarraute, E.; Costa, F. Generation of electrical energy for portable devices: Comparative study of an electromagnetic and a piezoelectric system. *Sens. Actuator A Phys.* **2004**,*116*, 461–471.
3. Roundy, S.; Wright, P.K.; Rabaey, J. A study of low level vibrations as a power source for wireless sensor nodes. *Comput. Commun.* **2003**, *26*, 1131–1144.
4. Pirisi, A.; Mussetta, M.; Gruosso, G.; Zich, R.E. Bio-Inspired Optimization Techniques for Wireless Energy Transfer. Proceedings of 2010 International Conference on Electromagnetics in Advanced Applications (ICEAA), Sydney, Australia, 20–24 September 2010; pp. 731–732.
5. Renaud, M.; Karakaya, K.; Sterken, T.; Fiorini, P.; Van Hoof, C.; Puers, R. Fabrication, modeling and characterization of MEMS piezoelectric vibration harvester. *Sens. Actuator A Phys.* **2008**,*145–146*, 380–386.
6. Jeon, Y.B.; Sood, R.; Jeong, J.; Kim, S.-G. MEMS power generator with transverse mode thin film PZT. *Sens. Actuator A Phys.* **2005**, *122*, 16–22.
7. Fang, H.B.; Liu, J.Q.; Xu, Z.Y.; Dong, L.; Wang, L.; Chen, D.; Cai, B.-C.; Liu, Y. Fabrication and performance of MEMS-based piezoelectric power generator for vibration energy harvesting.*Microelectron. J.* **2006**, *37*, 1280–1284.
8. Liu, J.Q.; Fang, H.B.; Xu, Z.Y.; Mao, X.-H.; Shen, X.-C.; Chen, D.; Liao, H.; Cai, B.-C. A MEMS-based piezoelectric power generator array for vibration energy harvesting. *Microelectron. J.* **2008**, *39*, 802–806.
9. Muralt, P.; Marzencki, M.; Belgacem, B.; Calame, F.; Basrour, S. Vibration energy harvesting with PZT micro device. *Proced. Chem.* **2009**, *1*, 1191–1194.
10. Defosseux, M.; Allain, M.; Defay, E.; Basrour, S. Highly efficient piezoelectric micro harvester for low level of acceleration fabricated with a CMOS compatible process. *Sens. Actuator A Phys.***2012**, *188*, 489–494.
11. Liu, H.; Quan, C.; Tay, C.J.; Kobayashi, T.; Lee, C. A MEMS-based piezoelectric cantilever patterned with PZT thin film array for harvesting energy from low frequency vibrations. *Phys. Proced.* **2011**, *19*, 129–133.

12. Gu, L. Low-frequency piezoelectric energy harvesting prototype suitable for the MEMS implementation. *Microelectron. J.* **2011**, *42*, 277–282.
13. Kim, M.; Hwang, B.; Ham, Y.H.; Jeong, J.; Min, N.K.; Kwon, K.H. Design, fabrication, and experimental demonstration of a piezoelectric cantilever for a low resonant frequency microelectromechanical system vibration energy harvester. *J. Micro/Nanolithogr. MEMS MOEMS* **2012**, *11*. [CrossRef]
14. Priya, S.; Inman, D.J. *Energy Harvesting Technologies*; Springer: New York, NY, USA, 2009.
15. Miller, L.M.; Halvorsen, E.; Dong, T.; Wright, P.K. Modeling and experimental verification of low-frequency MEMS energy harvesting from ambient vibrations. *J. Micromech. Microengi.* **2011**, *21*. [CrossRef]
16. Kok, S.L.; Rahman, A.; Fauzi, M. Design Considerations of MEMS Based Piezoelectric Cantilever for Harvesting Energy. Proceedings of 2012 IEEE Asia-Pacific Conference on Applied Electromagnetics (APACE), Melaka, Malaysia, 11–13 December 2012; pp. 238–242.
17. Andosca, R.; McDonald, T.; Genova, V.; Rosenberg, S.; Keating, J.; Benedixen, C.; Wu, J. Experimental and theoretical studies on MEMS piezoelectric vibrational energy harvesters with mass loading. *Sens. Actuat. A Phys.* **2012**, *178*, 76–87.
18. Wang, Q.; Pei, X.; Wang, Q.; Jiang, S. Finite element analysis of a unimorph cantilever for piezoelectric energy harvesting. *Int. J. Appl. Electromagn. Mech.* **2012**, *40*, 341–351.
19. Kim, M.; Hong, S.; Miller, D.J.; Dugundji, J.; Wardle, B.L. Size effect of flexible proof mass on the mechanical behavior of micron-scale cantilevers for energy harvesting applications. *Appl. Phys. Lett.* **2011**, *99*. [CrossRef]
20. Kasyap, A. Development of MEMS-Based Piezoelectric Cantilever Arrays for Vibrational Energy Harvesting. Ph.D. Thesis, University of Florida, Gainesville, FL, USA, 2007.
21. Shen, D.; Park, J.H.; Noh, J.H.; Choe, S.Y.; Kim, S.H.; Wikle, H.C., III; Kim, D.J. Micromachined PZT cantilever based on SOI structure for low frequency vibration energy harvesting. *Sens. Actuator A Phys.* **2009**, *154*, 103–108.
22. Kong, N.A.; Ha, D.S.; Erturk, A.; Inman, D.J. Resistive impedance matching circuit for piezoelectric energy harvesting. *J. Intell. Mater. Syst. Struct.* **2010**, *21*, 1293–1302.
23. Kim, S.B.; Park, H.; Kim, S.H. Comparison of MEMS PZT cantilevers based on *d31* and *d33*modes for vibration energy harvesting. *J. Microelectromech. Syst.* **2013**, *22*, 26–33.
24. Jambunathan, M.; Elfrink, R.; Vullers, R.; van Schaijk, R.; Dekkers, M.; Broekmaat, J. Pulsed Laser Deposited-PZT Based MEMS Energy Harvesting Devices. Proceedings of the 2012 International Symposium on Piezoresponse Force Microscopy and Nanoscale Phenomena in Polar Materials, Aveiro, Portugal, 9–13 July 2012; pp. 1–4.

25. Aktakka, E.E.; Peterson, R.L.; Najafi, K. Thinned-PZT on SOI Process and Design Optimization for Piezoelectric Inertial Energy Harvesting. Proceedings of the 2011 16th International IEEE Solid-State Sensors, Actuators and Microsystems Conference (TRANSDUCERS), Beijing, China, 5–9 June 2011; pp. 1649–1652.
26. Kim, M.; Lee, S.K.; Yang, Y.S.; Jeong, J.; Min, N.K.; Kwon, K.H. Design and fabrication of vibration based energy harvester using microelectromechanical system piezoelectric cantilever for low power applications. *J. Nanosci. Nanotechnol.* **2013**, *13*, 7932–7937.
27. Ottman, G.K.; Hofmann, H.F.; Lesieutre, G.A. Optimized piezoelectric energy harvesting circuit using step-down converter in discontinuous conduction mode. *IEEE Trans. Power Electron.* **2003**,*18*, 696–703.
28. Lefeuvre, E.; Audigier, D.; Richard, C.; Guyomar, D. Buck-boost converter for sensorless power optimization of piezoelectric energy harvester. *IEEE Trans. Power Electron.* **2007**, *22*, 2018–2025.
29. Kong, N.; Cochran, T.; Ha, D.S.; Lin, H.-C.; Inman, D.J. A Self-Powered Power Management Circuit for Energy Harvested by a Piezoelectric Cantilever. Proceedings of the 2010 25th Annual IEEE Applied Power Electronics Conference and Exposition (APEC), Palm Springs, CA, USA, 21–25 February 2010; pp. 2154–2160.

Chapter 11

A New Mirroring Circuit for Power MOS Current Sensing Highly Immune to EMI

Orazio Aiello * *and Franco Fiori*

Microelectronics EMC Group, Eln. Department, Politecnico di Torino, Corso Duca degli Abruzzi, 24, I-10129 Torino, Italy

ABSTRACT

This paper deals with the monitoring of power transistor current subjected to radio-frequency interference. In particular, a new current sensor with no connection to the power transistor drain and with improved performance with respect to the existing current-sensing schemes is presented. The operation of the above mentioned current sensor is discussed referring to time-domain computer simulations. The susceptibility of the proposed circuit to radio-frequency interference is evaluated through time-domain computer simulations and the results are compared with those obtained for a conventional integrated current sensor.

INTRODUCTION

In the last decades, the development of CMOS technologies in terms of scale integration and component performance has made the integration of full electronic system into a single chip possible. Such System-on-Chips (SoCs) comprise complex high-speed digital blocks, analog circuits and power section [1,2]. The power-supply voltage of such integrated circuits has decreased over time, leading to lower noise margins. However, the level of disturbances collected by the wiring harnesses of the electronic systems from the surrounding environment has strongly increased because of the widespread use of wireless systems for radio/TV broadcasting and mobile communications. In this context, the need to understand the effect of radio frequency interference (RFI) on baseband integrated circuits has continued to grow and several studies have tried to set out design guidelines and new circuit topologies for making integrated circuits immune to RFI [3–8]. Such disturbances superimposed onto nominal signals are of particular concern in the design of input and output front-end integrated circuits like those included in smart power ICs, which cannot be protected using EMI-filters at the PCB level because either their presence can affect the nominal circuit operation or the PCB is not present since the SoC is encapsulated inside the sensor (actuator) plastic holder. For instance, EMI filters and capacitors cannot be connected to the power transistor terminals because they would reduce the power efficiency. As a result, the power transistors that drive actuators through cables can experience high-level RFI that can lead to unexpected transistor switching and leakage currents, as shown in [9,10]. In addition to this problem, the circuits that are connected to the transistors' terminals, including circuits embedded in smart power ICs for monitoring the operation of the power transistors and the actuators, are also affected by RFI.

Within the aforementioned context, this paper investigates the susceptibility of integrated current sensors that are used in power circuits for control and/or safety purposes to RFI. The current of a power transistor can be sensed by using one of the circuits presented in [11], but only a few of them are suitable for integration on silicon. Among these integrated current sensors, one of the most frequently used is that based on a small transistor namely SenseFET. The voltages at the terminals of the SenseFET are kept equal to those of a power transistor in which the current to be detected flows. In this way, such a current is scaled by a constant K and provided by the SenseFET. K depends only on the aspect ratios (WL) of the two transistors. In this circuit, the SenseFET is electrically connected to the terminals of the power transistor and therefore it is affected by the RF disturbances collected by cables. In order to design embedded current sensor immune to such RF interferences, this paper presents a new integrated current-sensing solution conceived to be electrically not connected to the power transistor drain that can be affected by disturbances. The current is sensed whenever during its transient the power transistor and the SenseFET both operate in saturation. Therefore, the current flowing through the power transistor is mirrored by the SenseFET, is not dependent on the drain-source voltage and is only related with the mirroring factor K. The paper is organized as follows. Section 2 describes the operation of a common integrated current sensor. Section 3 summarizes the switching behavior of power MOSFETs. On the basis of such a behavior, a new current sensor not electrically connected to the drain terminal of the power transistor is proposed in Section 4. Then, the results of computer simulations aimed at evaluating the susceptibility of the proposed current sensor, the MagFET-based sensor and the commonly used current sensor to RFI are shown in Section 5. Finally, some concluding remarks are drawn in Section 6.

A CONVENTIONAL CURRENT SENSOR: THE SENSEFET

Common current sensors for integrated applications are based on the SenseFET circuit topology, in which the current $i_D(t)$ that flows through an MOS Power transistor is sensed using an elementary transistor of the same type that is used in the power transistor. Figure 1 shows the operation of the SenseFET current sensor, where the gate terminals of the aforementioned transistors are connected to each other and driven by an MOS driver, while the drain-to-source voltage of the SenseFET is made equal to that of the power transistor by means of a voltage follower. In this circuit, the relationship between the power transistor current (I_D) and the sensed current (I_{SENSE}) can be written as

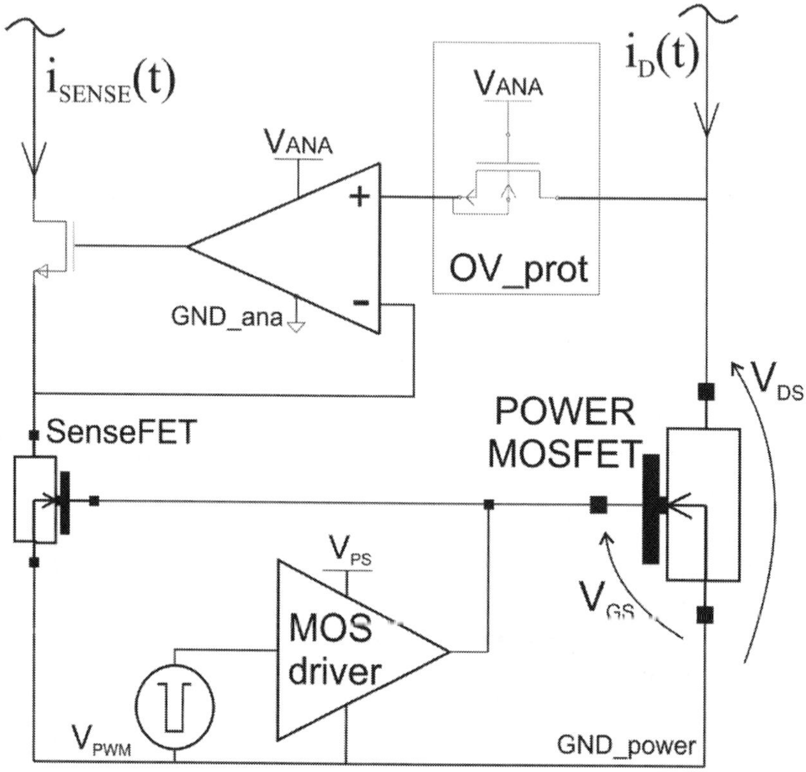

Figure 1. Current sensor based on the SenseFET technique.

$$K = \frac{\left(\frac{W}{L}\right)_{\text{power}}}{\left(\frac{W}{L}\right)_{\text{sense}}} \tag{1}$$

where (WL)power and (WL)sense are the aspect ratios of the power and SenseFET transistors, respectively. K is typically not lower than 10^3. The scaled current I_{SENSE} is then converted to a voltage, using a resistor to be used for digital signal processing. This paper focuses on a low side sensing topology highlighted in Figure 1.

However, the same reasoning can be applied for the high side topologies. In this technique the power dissipation is low and the accuracy is related to the matching of the SenseFET and the Power MOS. The two transistors are driven in the deep triode region by the same gate-source voltage and have to be kept at the same drain-source voltages. Therefore, to minimize the error in the sensed current, the amplifier needs to have a very high gain. Moreover, the amplifier should maintain the accuracy across the large input common-mode range. Although several circuit topologies of current sensors have been proposed, all of them include an amplifier, which implies stability issues, limited bandwidth and increased design complexity. In fact, the improvements in integrated current-sensing are basically focused on increasing amplifier performance [12–19]. Furthermore, the presence of the amplifier increases the overall susceptibility of the commonly employed SenseFET-based current sensing to EMI. In fact, this amplifier is connected through an over-voltage protection (see Figure 1) to the power transistor drain, and hence to the wiring interconnects at the system level that collect EMI. Since interference with an amplitude of hundreds of mV causes distortion in the amplifier and in the over-voltage protection, the operation of the current sensor can be impaired [6].

For all the above mentioned reasons, a new current sensor with improved performance due to the absence of any amplifier has been designed and its immunity to EMI is specifically addressed in this paper.

POWER MOSFET SWITCHING CHARACTERISTICS

The Power MOSFET is usually employed as a switch in circuits for energy conversion and management applications and to drive high-power loads. For instance, a Power MOSFET is used to control the current in inductive loads such as the windings of motors. A Power MOSFET that drives an inductive load is represented in Figure 2 where a freewheeling diode carries the load current when the Power MOS is switched-off and a stray inductance is included to account for package and board parasitic elements.

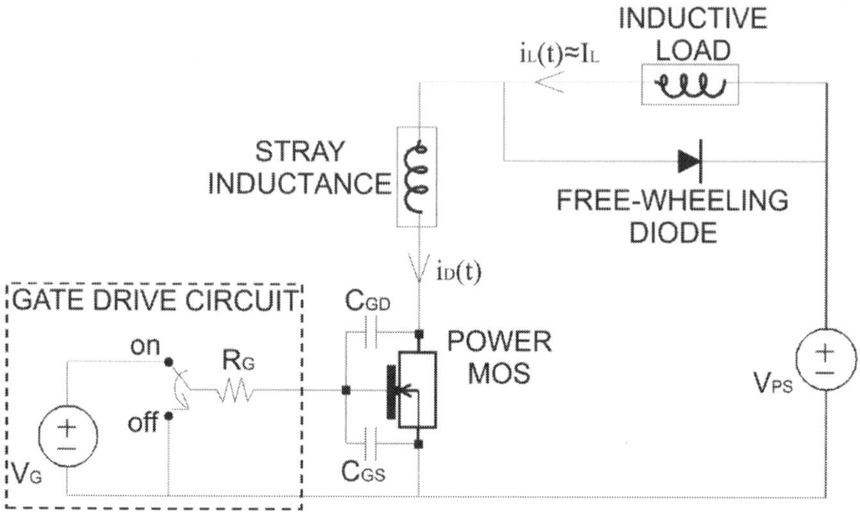

Figure 2. Power MOSFET device operating in an inductive load circuit.

The switching behavior of Power MOSFET structures is governed by the gate drive circuit and the nature of the load. The power transistor is switched on and off by a control or gate drive circuit, which can be represented (Thevenin's equivalent) as a DC voltage V_G with a series resistance R_G. The load current i_L transfers between the power MOSFET device and the freewheeling diode during each operating cycle. The inductor is charged and its current is increased when the Power MOSFET is turned-on while it is discharged when the load current flows through the diode. However, the change in the inductor current is small during one cycle, allowing the assumption that the current I_L is constant. On the basis of that, a new current sensor that operates during the transient of the Power MOSFET has been conceived. To this purpose, the waveform of the transient of a Power MOSFET is represented in Figure 3 and described in the following. Whenever the Power MOSFET is in off-state, the load current flows through the freewheeling diode. The initial conditions for the Power MOSFET are defined by $v_{GS}(0) = 0$, $i_D(0) = 0$,$v_{DS}(0) = V_{DSOFF}$. During the turn-on process, the gate bias voltage source V_G starts to charge the capacitances of the Power MOSFET. Since no drain current can flow through the power MOSFET device until the gate voltage exceeds its threshold voltage, the drain voltage initially remains at the drain bias voltage. The gate–drain capacitance $C_{GD}(V_{DSOFF})$ remains constant because the drain voltage is constant. Consequently, the time constant for charging the gate of the power MOSFET device is $R_G \cdot [C_{GS} + C_{GD}(V_{DSOFF})]$, resulting in a gate voltage given by and represented in Figure 3. The gate voltage reaches the threshold voltage at time t_1:

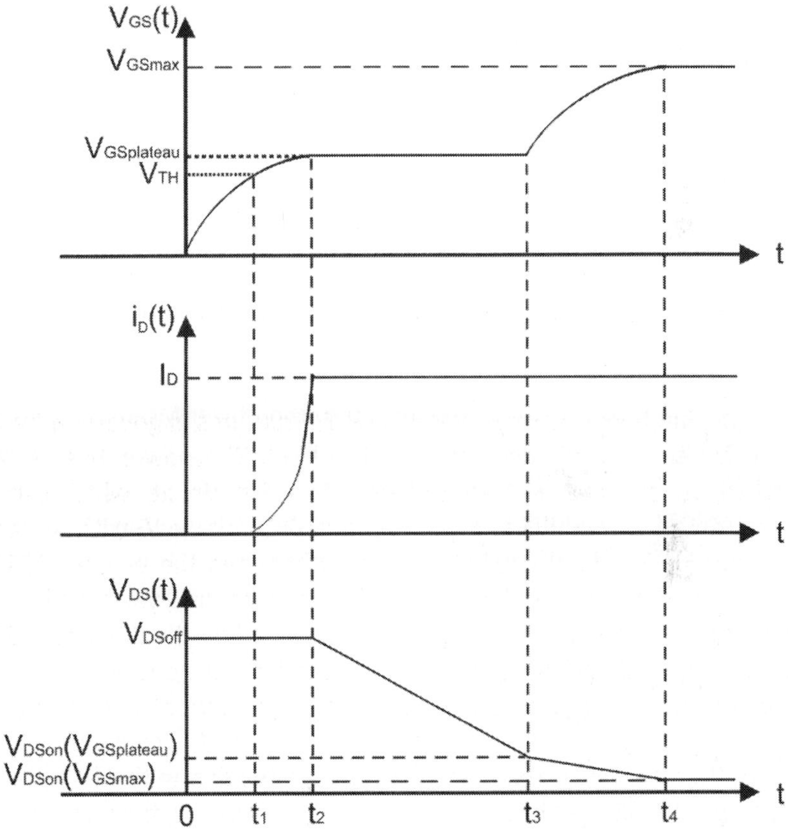

Figure 3. Waveforms for the power MOSFET during turn-on with a gate voltage source.

$$v_{GS}(t) = V_{GS_{max}} \cdot \left[1 - e^{-\frac{t}{R_G \cdot [C_{GS} + C_{GD}(V_{DS_{OFF}})]}}\right] \quad (2)$$

$$v_{\mathrm{GS}}(t)|_{\mathrm{t=t_1}} = V_{\mathrm{TH}}$$

(3)

$$t_1 = R_{\mathrm{G}} \cdot [C_{\mathrm{GS}} + C_{\mathrm{GD}}(V_{\mathrm{DS_{OFF}}})] \cdot ln\left(\frac{V_{\mathrm{GS_{max}}}}{V_{\mathrm{GS_{max}}} - V_{\mathrm{TH}}}\right) \quad (4)$$

Once the gate voltage exceeds the threshold voltage, drain current begins to flow.

$$i_{\mathrm{D}}(t) = \frac{\mu_{\mathrm{ni}} C_{\mathrm{OX}} W_{\mathrm{CH}}}{L_{\mathrm{CH}}} \cdot [v_{\mathrm{GS}}(t) - V_{\mathrm{TH}}]^2$$

(5)

Although the drain current increases, the drain voltage remains at the drain supply voltage V_{DSOFF}because the diode cannot sustain any voltage until all of the load current is transferred to the Power MOSFET. Since the drain voltage remains constant, the drain–gate capacitance is also invariant in the range $[t_1 - t_2]$. Consequently, the gate voltage continues to increase at an exponential rate as described byEquation (2) with the same time constant. The drain current increases as the square of the gate voltage as described by Equation (5) with a nonlinear waveform as represented in Figure 3.

The drain current increases until it becomes equal to the load current $I_D = I_L$ and $v_{GS}(t)$ reaches the voltage plateau $V_{GSplateau}$ at the time t_2.

$$v_{\mathrm{GS}}(t)|_{\mathrm{t=t_2}} = V_{\mathrm{GS_{plateau}}} = V_{\mathrm{GS_{max}}} \cdot \left[1 - e^{-\frac{t_2}{R_{\mathrm{G}} \cdot [C_{\mathrm{GS}} + C_{\mathrm{GD}}(V_{\mathrm{DS_{OFF}}})]}}\right]$$

(6)

$$t_2 = R_{\mathrm{G}} \cdot [C_{\mathrm{GS}} + C_{\mathrm{GD}}(V_{\mathrm{DS_{OFF}}})] \cdot ln\left(\frac{V_{\mathrm{GS_{max}}}}{V_{\mathrm{GS_{max}}} - \sqrt{\frac{I_{\mathrm{D}} L_{\mathrm{CH}}}{\mu_{\mathrm{n}} C_{\mathrm{ox}} W_{\mathrm{CH}}}} - V_{\mathrm{TH}}}\right) \quad (7)$$

All of the load current has transferred from the diode to the Power MOSFET device at time t_2 and the diode is now able to support voltage. The drain-source voltage of the Power MOSFET starts to reduce at this time. Since the drain current is constant and equal to the load current ($I_D = I_L$), the gate voltage at time t_2 can be also expressed as

$$v_{GS}(t)|_{t_2-t_3} = V_{GS_{plateau}} = V_{TH} + \sqrt{\frac{I_D L_{CH}}{\mu_n C_{ox} W_{CH}}}. \quad (8)$$

The gate-source voltage remains constant at the plateau voltage until the drain-source voltage has reduced to the on-state voltage drop corresponding to the product of the load current and the on-resistance of the device at a gate bias equal to the plateau voltage. Since the gate voltage is constant during the plateau phase, all the gate current $i_{Gplateau}$ is used to charge the gate–drain or Miller capacitance. The gate current during the plateau phase is given by

$$i_{G_{plateau}} = \frac{V_{GS_{max}} - V_{GS_{plateau}}}{R_G}. \quad (9)$$

As this current charges the gate–drain capacitance, its voltage decreases at a rate given by

$$\frac{dv_{GD}(t)}{dt} = -\frac{i_{GP}(t)}{C_{GD}(v_{DS}(t))}. \quad (10)$$

Since the gate–source voltage is constant at $V_{GSplateau}$ during this time, the drain voltage also decreases linearly with time:

$$\frac{dv_{DS}(t)}{dt} = \frac{dv_{GD}(t)}{dt} = -\frac{i_{GP}(t)}{C_{GD}(v_{DS}(t))} = -\frac{V_{GS_{max}} - V_{GS_{plateau}}}{R_G \cdot C_{GD}(v_{DS}(t))}. \quad (11)$$

At the end of the plateau phase at time t_3, the drain–source voltage becomes equal to the on-state voltage drop corresponding to the plateau gate bias voltage. Based on this, the duration of the plateau can be expressed as:

$$\int_{t_2}^{t_3} dv_{\mathrm{D}} = \int_{t_2}^{t_3} -\frac{V_{\mathrm{GS}_{\mathrm{max}}} - V_{\mathrm{GS}_{\mathrm{plateau}}}}{R_{\mathrm{G}} \cdot C_{\mathrm{GD}}(v_{\mathrm{DS}}(t))} dt \tag{12}$$

$$t_3 - t_2 = \frac{R_{\mathrm{G}} \cdot C_{\mathrm{GD}_{\mathrm{AVG}}}}{V_{\mathrm{GS}_{\mathrm{max}}} - V_{\mathrm{GS}_{\mathrm{plateau}}}} \cdot \left[V_{\mathrm{DS}_{\mathrm{OFF}}} - I_{\mathrm{D}} R_{\mathrm{DS}_{\mathrm{ON}}}(V_{\mathrm{GS}_{\mathrm{plateau}}})\right] \tag{13}$$

where C_{GDav} is the gate–drain capacitance assumed to have a constant average value during the transient and $R_{DSON}(V_{GSplateau})$ is the on-resistance of the power MOSFET device for a gate-source voltage equal to the plateau voltage. Beyond the plateau, the gate voltage increases exponentially again as shown in theFigure 3 until it reaches the gate supply voltage. The time constant for this exponential rise is different from the initial phase due to the large gate–drain capacitance. The increasing gate voltage produces a reduction of the on-resistance of the power MOSFET device, resulting in a small reduction of the drain voltage during this fourth phase of the turn-on process.

Considering reasonable value for the parasitics of the Power transistor sized in the order of mm^2 and for the integrated driving resistance R_G, the interval time ($t_3 - t_2$) results to be larger than hundreds of nanoseconds. On the basis of a trigger event occurring during this interval, a new current sensor is proposed in Section 4.

A NEW CURRENT SENSOR BASED ON THE MILLER EFFECT

In order to design a current sensor immune to EMI, a new circuit is conceived, which is not electrically connected with the Power MOSFET drain terminal that is prone to disturbances. To this purpose, the current that flows through the Power MOSFET (i_D) is mirrored and processed during the V_{GS} voltage plateau due to the Miller effect (between points 2 and 3 in Figure 3). In this region, the Power MOSFET operates in the saturation region and the mirrored current I_{MIRROR} is not dependent on the drain-source voltage and is only related with the mirroring factor K. The Power MOSFET is driven by a PWM signal that provides both the proper-timing driving voltage and a negative pulse that generates a switching transient with negligible reduction of the current in inductive loads.

The block diagram of the proposed current sensor is shown in Figure 4. A low-pass RC filter (R_F– C_F) with a cut-off frequency significantly lower than the frequency of the EMI disturbances is placed between the Power MOSFET and SenseFET gates. In particular, $R_F = 10\ k\Omega$ and $C_F = 2\ pF$, so that the cut-off frequency is less than 10 MHz ($f_{CUT\_OFF} \approx 8\ MHz$). The current $I_{unknown}$ is mirrored by a SenseFET, processed and converted in a serial code (Analog to Digital Converter in Figure 4, further reported inFigure 5). The SenseFET current, in turn, is mirrored 128 times. Each of these currents is compared with progressively scaled current references in order to provide a thermometric digital code of the SenseFET current I_{MIRROR}. Then, a further thermometric-to-binary conversion is provided [20]. As soon as a trigger event occurs, the binary code is written in a register. Such a binary word provides the value of the current to detect $I_{unknown}$, which is then converted to a serial code.

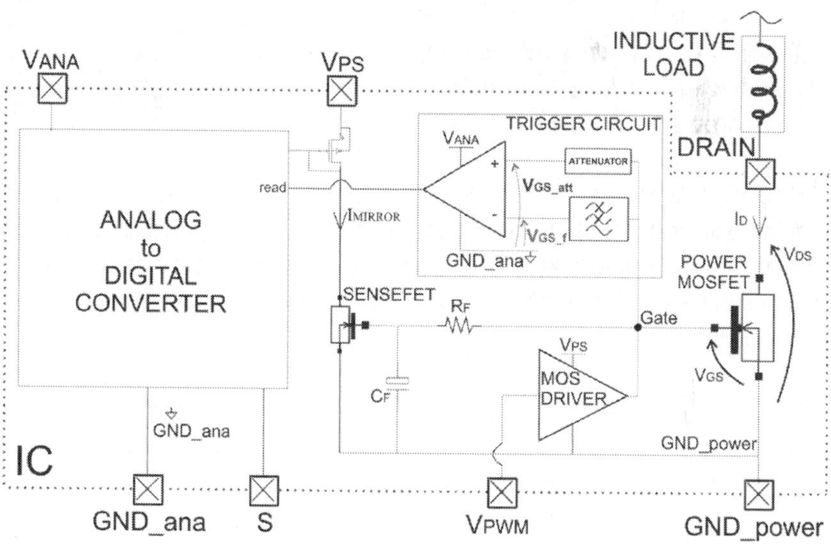

Figure 4. Block diagram of the proposed current sensor that exploits the Miller effect.

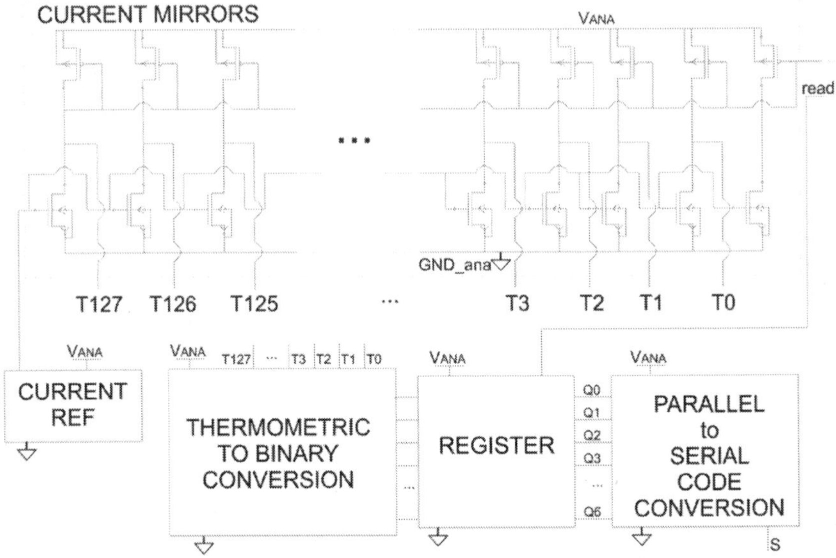

Figure 5. Analog to Digital Converter in Figure 4.

The trigger circuit is based on the comparison of an attenuated gate-source characteristic V_{GSatt} with the gate-source voltage after a low-pass filtering process V_{GS_f}. Both these two voltages are obtained from the gate-source voltage V_{GS} by means of only passive component as represented in Figure 6. An inverting stage made of a high voltage inverter allows to null the current dissipation whenever the V_{GS}voltage transient ends. Neglecting the drop voltage across the transistor M_{SD} that operates in triode during the transient, the voltage V_{GS_att} is given by:

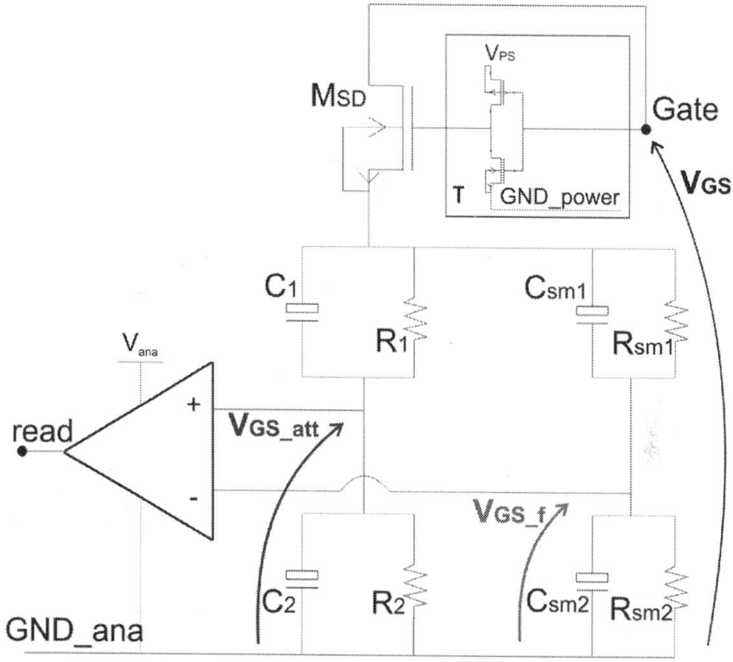

Figure 6. Trigger circuit.

$$V_{GS\_att} = \frac{R_2(1+sR_1C_1)}{R_2(1+sR_1C_1)+R_1(1+sR_2C_2)}V_{GS} \tag{14}$$

Setting $R_1C_1 = R_2C_2$ the voltage V_{GSatt} is simply the gate-source voltage V_{GS} scaled by a factork_{din_adp} independently by the frequency components of the switching transient of the Power MOSFET.

$$V_{GS\_att} = \frac{R_2}{R_1+R_2}V_{GS} = \frac{C_1}{C_1+C_2}V_{GS} = k_{din\_adp} \cdot V_{GS} \tag{15}$$

where k_{din_adp} is set in order to not overcome the maximum voltage deliverable to the analog blocks.

$$k_{din\_adp} = \frac{R_2}{R_1 + R_2} = \frac{C_1}{C_1 + C_2} \tag{16}$$

The voltage V_{GS_f} is obtained by a filtering and the passive components (R_{sm1}, C_{sm1}, R_{sm2} and C_{sm2}) are set to makes this voltage V_{GS_f} lower than the attenuated voltage V_{GSatt} during the interval $[0 - t_1]$. At the same time, the passive components are set such that the filtered voltage V_{GS_f} reaches the attenuated voltage V_{GSatt} at the time t^* during the plateau, as sketched in Figure 7.

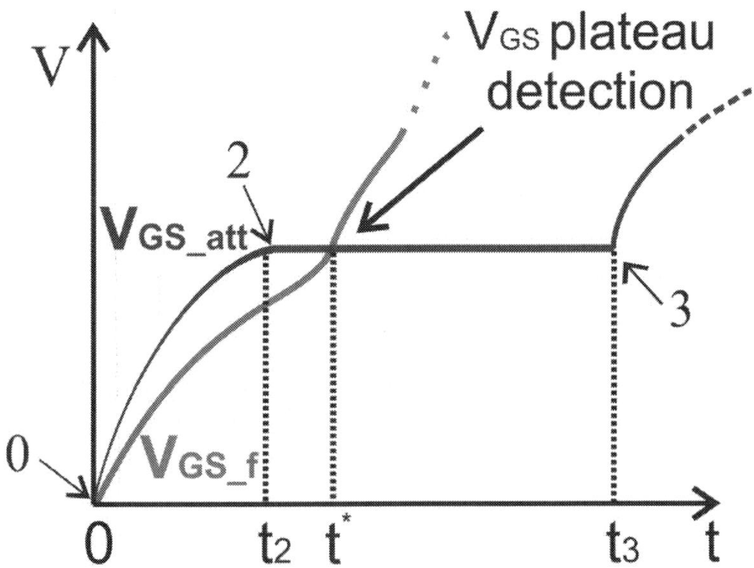

Figure 7. Principle of the detection of the gate-source plateau.

$$\begin{cases} V_{GS\_f}(t) < V_{GS\_att}(t) & \text{if } 0 < t < t^* \\ V_{GS\_f}(t) > V_{GS\_att}(t) & \text{if } t^* < t < t_3 \end{cases} \tag{17}$$

Assuming V_{GS} as a ramp before the voltage plateau, it can be shown that the filtered voltage $V_{GS\_f}(t)$ can be expressed as:

$$V_{GS\_f}(t) = \begin{cases} \frac{mG}{(s_p)^2}(s_p - s_z)(1 - e^{-s_p \cdot t}) + \frac{mGs_z}{s_p} \cdot t & \text{if } 0 < t < t_2 \\ \frac{mG}{s_p^2}(s_p - s_z)e^{-s_p \cdot t}(e^{s_p \cdot t_2} - 1) + \frac{mGs_z}{s_p} \cdot t_2 & \text{if } t_2 < t < t_3 \end{cases} \tag{18}$$

where

$$m = \frac{V_{TH}}{t_1} \tag{19}$$

$$G = \frac{C_{sm1}}{C_{sm1} + C_{sm2}} \tag{20}$$

$$s_z = \frac{1}{R_{sm1} \cdot C_{sm1}} \tag{21}$$

$$s_p = \frac{R_{sm1} + R_{sm2}}{R_{sm1} \cdot R_{sm2} \cdot (C_{sm1} + C_{sm2})} \tag{22}$$

On the basis of the two considered voltages V_{GS_f} and V_{GS_att}, it is possible to find out the time t^* during the voltage plateau when they are equal.

$$V_{GS\_f}(t^*)|_{(t_2 < t^* < t_3)} = k_{din\_adp} V_{GS_{plateau}} \tag{23}$$

$$\frac{mG}{s_p^2}(s_p - s_z)e^{-s_p \cdot t^*}(e^{s_p \cdot t_2} - 1) + \frac{mGs_z}{s_p} \cdot t_2 = k_{din\_adp} V_{GS_{plateau}} \tag{24}$$

$$t^* = -\frac{1}{s_p} \cdot ln \left[\frac{k_{din\_adp} V_{GS_{plateau}} - \frac{mGs_z}{s_p} t_2}{\frac{G}{s_p^2}(s_p - s_z)(e^{s_p \cdot t_2} - 1)} \right]. \tag{25}$$

Firstly, a minimum value for C_{sm1} = 200 fF is set and a reasonable value of R_{sm1} = 62 *kΩ* is chosen to set the zero s_z = 12.8 *MHz*. Consequently, the value of C_{sm2} = 2.5 *pF* and R_{sm2} = 19 *kΩ* are set in order to place the pole s_p = 4.05 *MHz* before the zero. In this way an instant t^* in the range [$t_2 - t_3$] according toEquation (25) is found out and a correct trigger event can be provided. As soon as the mentioned two voltages V_{GS_att} and V_{GS_f} reach the same value during the gate-source plateau voltage due to the Miller effect, a comparator finds out a trigger event. Such an event enables to write the register whenever the Power MOSFET is in saturation during its switching transient. In the register, the digital value of the SenseFET current I_{MIRROR} is stored. Such a value represents the current to sense $I_{unknown}$ scaled by the mirroring ratio K.

The proposed circuit is suitable for inductor current monitoring in DC-DC converter that operates in Continuous Conduction Mode (CCM) and whenever the detection of the current flowing in a Power MOSFETs operating in PWM that drive resistive and inductive loads is needed. The current to be monitored ranges from 100 μA to 5 A, and the mirroring factor K has been set **K** = 10^4. The sensibility of the designed sensor is 50 μA (LSB). The absence of the amplifier provides an higher immunity to EMI and a faster response in inductor-current monitoring than the traditional current-sensing technique as shown in Section 5.

PREDICTION OF INTEGRATED CURRENT SENSORS TO RFI

This section shows the results of the time-domain simulations carried out to evaluate the susceptibility of the proposed current sensor and the traditional sensor to RFI.

The proposed new current sensor is based on the transient behavior of the Power MOSFET in which the current to detect flows, and such a transient strongly depends on the parasitics of the Power transistor itself. For this reason, no other additional parasitic capacitances due to the test circuit have to affect the transient behavior of the Power MOSFET. Therefore, a CW RFI current of magnitude i_{RF} = 200 mA has been superimposed onto the current to detect I_D by means of a toroidal RF transformer.

In fact, the analyses of the susceptibility to RFI have been carried out referring to the schematic view ofFigure 8 similarly to immunity tests [21]. A resistor R_{BIAS} sets the value of the current I_D.

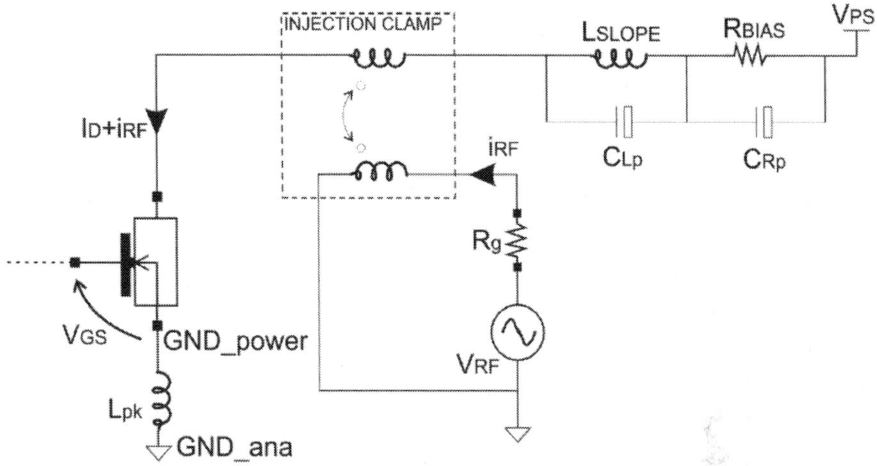

Figure 8. Schematic view of the Power MOSFET considered in the time-domain simulations.

Figure 9 shows the switching transient traces (V_{DS}, V_{GS}) of a Power MOSFET in which the current to be detected I_D flows, as well as the respective currents provided by a traditional SenseFET sensor (I_{SENSE}) and the proposed new sensor (I_{MIRROR}). In such a time-domain simulations, a CW RFI current i_{RF} = 200 *mA* at 100 MHz superimposed on the drain current have been considered. The currents provided by the traditional SenseFET sensor (I_{SENSE}) and by the proposed new sensor (I_{MIRROR}) versus current to sense I_D and their minimum and maximum errors due to the aforementioned RFI magnitude at 10 MHz, 100 MHz and 400 MHz are shown respectively in Figures 10–12. Furthermore, the minimum and maximum errors of the two considered currents due to a CW RFI i_{RF} = 200 *mA* superimposed on the drain current I_D versus frequency are reported in Figure 13. The results highlight that the immunity of the proposed current sensor to RFI is significantly greater than that of the conventional SenseFET circuit shown in Figure 1.

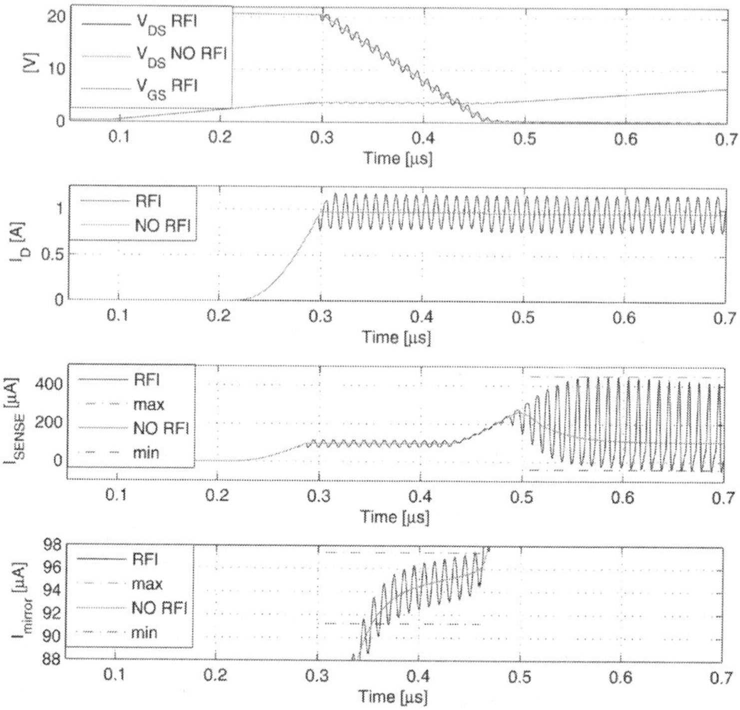

Figure 9. Switching transient traces of a Power MOSFET (V_{DS},V_{GS}, I_D) and the respective currents provided by a traditional SenseFET sensor (I_{SENSE}) and the new sensor based on the Miller effect (I_{MIRROR}). CW RFI current amplitude 200 mA @ 100 MHz superimposed on the drain current.

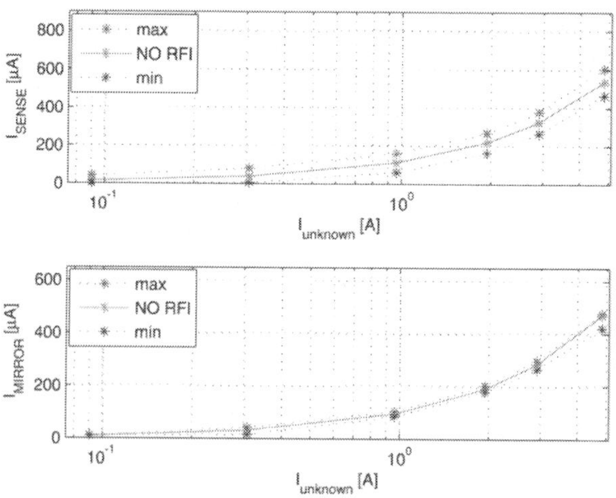

Figure 10. Maximum and minimum current-sensing of the traditional SenseFET current sensor (I_{SENSE}) and the new sensor based on the Miller effect (I_{MIRROR}) *versus* current to sense I_D. CW RFI current amplitude 200 mA @ 10 MHz.

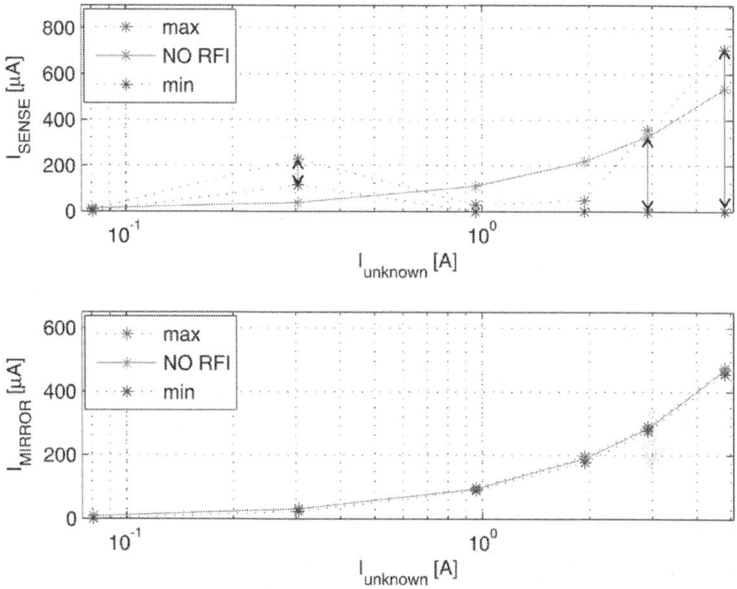

Figure 11. Maximum and minimum current-sensing of the traditional SenseFET current sensor (I_{SENSE}) and the new sensor based on the Miller effect (I_{MIRROR}) *versus* current to sense I_D. CW RFI current amplitude 200 mA @ 100 MHz.

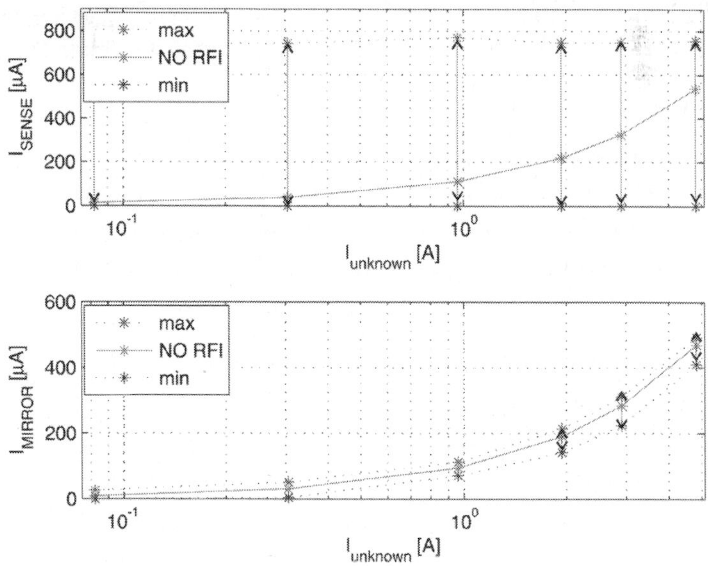

Figure 12. Maximum and minimum current-sensing of the traditional SenseFET current sensor (I_{SENSE}) and the new sensor based on the Miller effect (I_{MIRROR}) *versus* current to sense I_D. CW RFI current amplitude 200 mA @ 400 MHz.

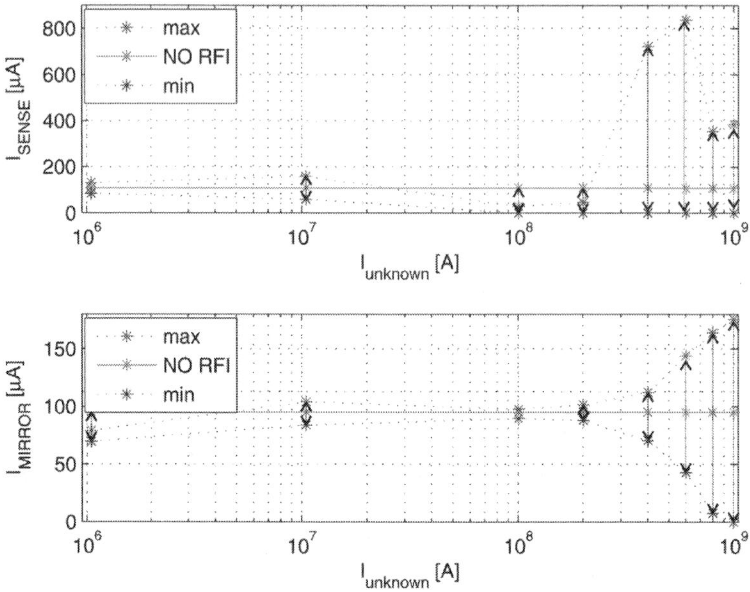

Figure 13. Maximum and minimum current-sensing of the traditional SenseFET current sensor (I_{SENSE}) and the new sensor based on the Miller effect (I_{MIRROR}) *versus* frequency. CW RFI current amplitude 200 mA superimposed on a currentI_D = 1A.

CONCLUSIONS

In this work, the susceptibility of common integrated sensors to RFI has been discussed for power transistor current monitoring. In order to improve the immunity of Power MOSFET transistor to RFI in current sensing, a new integrated solution has been proposed. It exploits the Miller effect on the switching transient of the Power MOSFET. The detection of the gate-source voltage plateau due to the Miller effect allows to mirror the current to be detected whenever the Power MOSFET operates in saturation. In this way, the mirrored current is not dependent on the drain-source voltage and the RFI that reaches the drain terminal of the Power MOSFET does not strongly influence the operation of the current sensor. Furthermore, there is no need to connect any amplifier to the drain terminal of the Power MOSFET, hence eliminating a barrier on the overall current sensor performance. The operation of the new current sensing circuit and that of a conventional current sensor have been compared in time-domain simulations. Since the novel sensor operates during the switching transient, it provides a faster current detection than the traditional sensor. Whereas the amplifier usually limits the performance of the SenseFET-technique, no such issue has to be considered for the new sensor. The susceptibility of the proposed current monitoring solution has been evaluated through time-domain simulations and compared with that of a conventional current sensor. The analyses carried out in this work have shown a significant improvement in the immunity to RFI in the range 1 MHz–1 GHz that the new proposed solution enables.

REFERENCES

1. Murari, B.; Bertotti, F. *Smart Power ICs*; Springer: Berlin, Germany, 1995.
2. Baliga, B.J. *Fundamentals of Power Semiconductor Devices*; Springer: New York, NY, USA, 2008.
3. Richelli, A. Increasing EMI immunity in novel low-voltage CMOS OpAmps. *IEEE Trans. EMC***2012**, *54*, 947–950.
4. Fiori, F. Susceptibility of CMOS voltage comparators to radio frequency interference. *IEEE Trans. EMC* **2012**, *54*, 434–442.
5. Aiello, O.; Fiori, F. On the susceptibility of embedded thermal shutdown circuit to radio frequency interference. *IEEE Trans. EMC* **2012**, *54*, 405–412.
6. Fiori, F.; Crovetti, P.S. Complementary differential pair with high immunity to RFI. *IEEE Electron. Lett.* **2002**, *38*, 1663–1664.
7. Redoute, J.M.; Steyaert, M. *EMC of Analog Integrated Circuits*; Springer: New York, NY, USA; January; 2010.
8. Ramdani, M.; Sicard, E.; Boyer, A.; Ben Dhia, S.; Whalen, J.J.; Hubing, T.H.; Coenen, M.; Wada, O. The electromagnetic compatibility of integrated circuits—Past, present and future. *IEEE Trans. EMC* **2009**, *51*, 78–100.
9. Bona, C.; Fiori, F.L. A new filtering technique that makes power transistors immune to EMI. *IEEE Trans. Power Electron.* **2011**, *26*, 2946–2955.
10. Jovic, O.; Stuermer, U.; Wilkening, W.; Maier, C.; Baric, A. Susceptibility of PMOS Transistors under High RF Excitations at Source Pin. Proceedings of the 20th International Zurich Symposium, Zurich, Swiss, 12–16 January 2009; Volume 58. pp. 401–404.
11. Patel, A.; Ferdowsi, M. Current sensing for automotive electronics? A survey. *IEEE Trans. Veh. Technol.* **2009**, *58*, 4108–4119.
12. Fai, L.C.; Mok, P.K.T. A monolithic current-mode CMOS DC-DC converter with on-chip current-sensing technique. *IEEE J. Solid State Circ.* **2004**, *39*, 3–14.
13. Chen, J.J.; Su, J.H.; Lin, H.Y.; Chang, C.C.; Lee, Y.; Chen, T.C.; Wang, H.C.; Chang, K.S.; Lin, P.S. Integrated current sensing circuits suitable for step-down DC-DC converters. *IEEE Electron. Lett.* **2004**, *40*, 200–202.

14. Leung, C.Y.; Mok, P.K.T.; Leung, K.N.; Chan, M. An integrated CMOS current-sensing circuit for low-voltage current-mode buck regulator. *IEEE Trans. Circ. Syst.* **2005**, *52*, 394–397.
15. Dake, T.; Ozalevli, E. A precision high-voltage current sensing circuit. *IEEE Trans. Circ. Syst.* **2008**, *55*, 1197–1202.
16. Yuan, B.; Lai, X. On-chip CMOS current-sensing circuit for DC-DC buck converter. *IEEE Electron. Lett.* **2009**, *45*, 102–103.
17. Mengmeng, D.; Hoi, L. An integrated speed- and accuracy-enhanced CMOS current sensor with dynamically biased shunt feedback for current-mode buck regulators. *IEEE Trans. Circ. Syst.* **2010**, *57*, 2804–2814.
18. Mengmeng, D.; Hoi, L.; Jin, L. A 5-MHz 91% peak-power-efficiency buck regulator with auto-selectable peak- and valley-current control. *IEEE J. Solid State Circ.* **2011**, *46*, 1928–1939.
19. Leung, C.Y.; Mok, P.K.T.; Leung, K.N. A 1-V integrated current-mode boost converter in standard 3.3/5-V CMOS technologies. *IEEE J. Solid State Circ.* **2005**, *40*, 2265–2274.
20. Sail, E.; Vesterbacka, M. Thermometer-to-Binary Decoders for Flash Analog-to-Digital Converters. Proceedings of the 18th European Conference on Circuit Theory and Design, Seville, Spain, 26–30 August 2007; pp. 240–243.
21. *Road Vehicles: Component Test Method for Electrical Disturbances from Narrowband Radiated Electromagnetic Energy. Part 4: Bulk Current Injection (BCI)*; The International standard ISO 11452-4; ARRB Group Limited: Vermont South, Australia, 2001.

CITATION

2013 by the authors; licensee MDPI, Basel, Switzerland. This article is an open access article distributed under the terms and conditions of the Creative Commons Attribution license (http://creativecommons.org/licenses/by/3.0/).

Chapter 12

Effects of Germanium Incorporation on Optical Performances of Silicon Germanium Passive Devices for Group-IV Photonic Integrated Circuits

Seongjae Cho[1], *Joonsuk Park*[2], *Hyungjin Kim*[3], *Robert Sinclair*[2], *Byung-Gook Park*[3,], *James S. Harris Jr.*[1]

[1] Department of Electrical Engineering, Stanford University, Stanford, CA 94305, USA
[2] Department of Materials Science and Engineering, Stanford University, Stanford, CA 94305, USA
[3] Inter-university Semiconductor Research Center (ISRC) with Department of Electrical Engineering and Computer Science, Seoul National University, Seoul 151-742, Republic of Korea

Highlights

- Evaluation of optical performances of $Si_{1-x}Ge_x$ waveguide and microring resonator by optical simulation.
- Establishment of permissible ranges of Ge fraction for the SiGe passive components.
- Refractive index (*n*) and extinction coefficient (*k*) have been mathematically modeled in a closed-form function of Ge fraction at 1550-nm wavelength.
- Investigation of germanium morphology in various materials systems by HR-TEM.

ABSTRACT

Optical interconnect in integrated optoelectronic circuits is one of the promising next-generation technologies for replacing metalized interconnect. Efforts have been made to use silicon (Si)-compatible materials such as germanium (Ge) and Ge-buffered III–V compound semiconductors, along with Si, as optical sources for Si and group-IV integrated optoelectronic systems. This opens the possibility that higher fraction of Ge with its high refractive index (*n*) can be incorporated in Si waveguide for optical interconnect and the graftability between Si and group-IV or III–V materials would be improved in silicon photonics. In this work, advantageous features of nano-structured silicon germanium ($Si_{1-x}Ge_x$) optical waveguide with different Ge fraction (*x*) were evaluated by both optical simulations and theoretical calculations, which are mainly found in the enhanced optical confinement and better interfacing capability. Along with the SiGe waveguide, performance of $Si_{1-x}Ge_x$ microring resonator under material loss in the effect of extinction coefficient (*k*) has been investigated to suggest the necessity of optimizing the Ge content in $Si_{1-x}Ge_x$ passive devices. While carrying out the establish design criteria, *n* and *k* have been modelled in closed-form functions of Ge fraction at 1550 nm. Furthermore, by examining high-resolution transmission electron microscopy (HR-TEM) images, process compatibility of Ge with either group-IV alloys or III–V compound semiconductors is confirmed for the monolithically integrated photonic circuits.

INTRODUCTION

Optical interconnect is considered one of the emerging technologies for replacing copper (Cu), particularly, in its application to chip-level (on-chip and chip-to-chip) interconnect, owing to its genuine virtues of high bandwidth, low power, low latency, and noise immunity [1]. The most attractive aspects of fabricating waveguides from Si are the low initial cost of the material; the mature and well-characterised processing techniques that are underpinned by decades of research, development, and manufacturing in the microelectronics industry; and the potential for straightforward integration with electronic devices on the same substrate [2]. The majority of silicon (Si) waveguides for Si photonics have been fabricated on the silicon-on-insulator (SOI) platform. However, Si has internal quantum efficiency as low as the order of 10^{-5} which restricts its potential as an efficient optical source in Si photonics. Although there have been attempts to achieve lasing from Si in many ways, the processing is quite complicated, and the optical efficiency still has much room to improve [3], [4] and [5]. To address this concern, opportunities have been found by developing optical sources using Si-compatible materials such as germanium (Ge) or even III–V compound semiconductors on Si with a Ge buffer [6], [7], [8], [9] and [10]. Under these circumstances, Ge can be incorporated for constructing SiGe waveguide and other passive devices owing to its intermediate lattice constant and higher optical confinement for Si and group-IV integrated photonic circuits. Although several merits are definitely expected by incorporating Ge, there are competing demerits in device performances, as summarized in Fig. 1. Optical confinement is one of the most crucial requirements for waveguiding material, for example, increased material loss by higher refractive index (n) should be considered simultaneously in designing the passive devices.

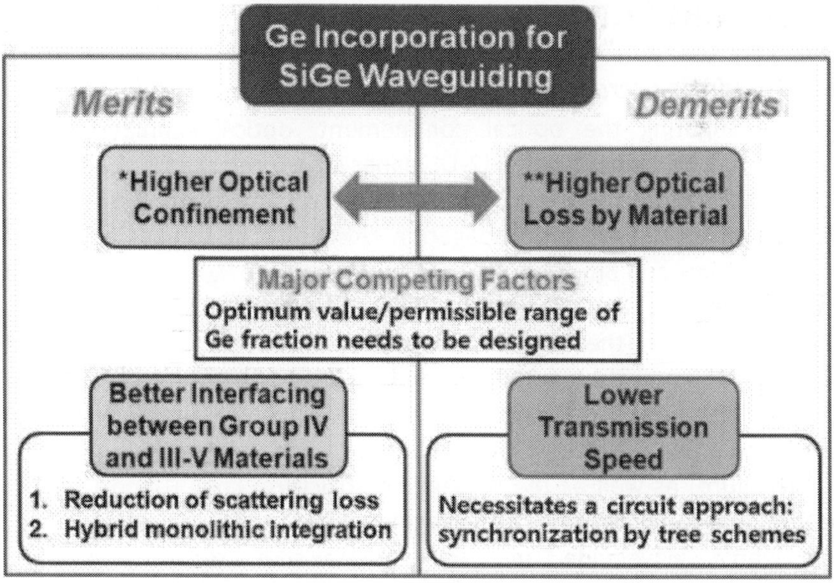

Figure 1. Effects of Ge incorporation on the performances of SiGe passive optical components.

In this work, we evaluated the optical confinement of $Si_{1-x}Ge_x$ waveguides with different Ge fractions (x), time delays caused by introducing higher-n material, and permissible range of x in consideration of extinction coefficient (k) in order to confirm the validity of $Si_{1-x}Ge_x$ as a waveguiding material for on-chip interconnect by a series of optical simulations and theoretical calculations. Optical cavity performances including full wave width at half maximum (FWHM) and quality factor (Q-factor) of $Si_{1-x}Ge_x$ microring resonator are also investigated along with the effect n and k, by which $x = 0.4–0.6$ comes as a permissible range of Ge content. While carrying out the evaluations, n and k have been mathematically modelled in a closed-form function of x at 1550-nm wavelength light signal, which is another genuine task performed in this work. Furthermore, analysis images from Ge interfaces were provided by high-resolution transmission electron microscopy (HR-TEM) to ensure its process viability in various material systems and confirm surface roughness within the permissible range to obtain interface scattering loss below a magnitude of 10^{-2} dB/mm order at sub-micron dimension.

2. BACKGROUNDS

2.1. Simulation strategy

For investigating the optical confinement, optical simulations were carried out in a 2-dimensional (2-D) planar structure [11]. Fig. 2 shows the schematic of a simulated waveguide structure with its plane is perpendicular to the light propagation direction. The waveguide consists of Si (bottom) and $Si_{1-x}Ge_x$ (top) slabs. W_{WG}, $H_{WG,Si}$, $H_{WG,SiGe}$ are the width of the waveguide, and the heights of Si and SiGe slabs, respectively. In the simulations, these dimensions were set to be $W_{WG} = 450$ nm, $H_{WG,Si} = 250$ nm, and $H_{WG,SiGe} = 500$ nm. Line A indicates the interface between Si and $Si_{1-x}Ge_x$ and the geometrical centre of the simulated domain including the silicon dioxide (SiO_2) cladding (not shown in the figure) as well as the waveguide structure; lines B and C are the bottom and top boundaries of waveguide; line D and E indicate centres of the whole waveguide and the $Si_{1-x}Ge_x$ slab, respectively. These distinguishable lines allow for faster tracing of peaks in the optical profiles. As will be clearer in a following section, the Si slab was intentionally introduced to investigate how cohesively the $Si_{1-x}Ge_x$ slab drags an optical mode towards its centre from the Si slab as x increases. Considering the Si slab simultaneously provides useful results from a practical view because it should be more strategic and cost-effective to grow a $Si_{1-x}Ge_x$ waveguide epitaxially on Si platform, either a Si or extremely thin silicon-on-insulator (SOI) platform than to use a Ge platform. However, in a later section, $Si_{1-x}Ge_x$ waveguides with identical dimension (W_{HG} = 450 nm, H_{WG} = 500 nm) without Si slab were simulated to make a fair comparison of the spatial mode profiles and optical confinements inside the structures with a genuine material solely determined by the value of x. Later in the simulation part, mathematical approaches are rigorously made for extracting the permissible ranges of Ge fraction for SiGe waveguide with a unit length of 100 μm in terms of half compensated transmission power (CPT) (maximum Ge fraction at 3-dB cut-off) and for SiGe microring resonator to appreciate the effects of Ge incorporation from a more practical viewpoint.

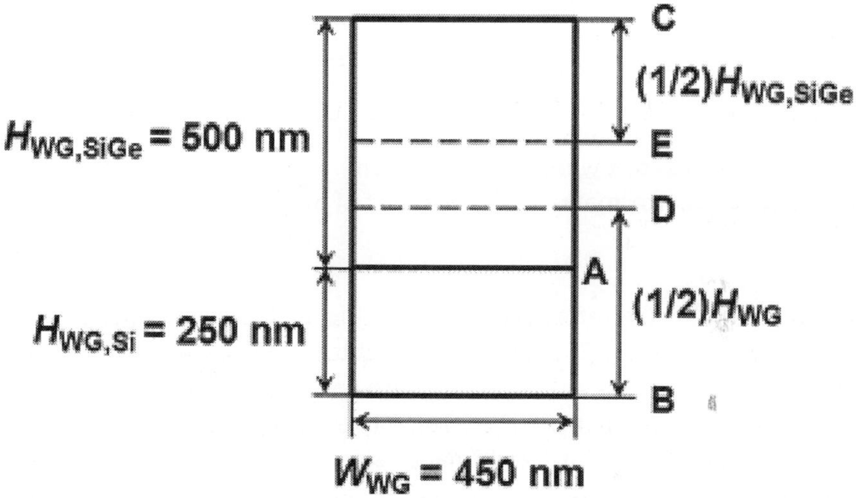

Figure 2. Schematic view of the simulated waveguide plane.

2.2. Physical parameters

The wavelength of the incident light normal to the plane was 1550 nm where the optical attenuation in the silica-based optical fibre is minimized [12]. The optimum wavelength for chip-level optical interconnect is not a critical issue, since the length of light propagation is too short to be of concern considering the attenuation reliability of long-distance communication. However, by adopting an optical source with a 1550-nm wavelength and corresponding optical components, a chip can be more readily evolved into a sub-system in a higher-scale integrated optoelectronic system of which individual subordinate chips are linked by external optical fibres. This wavelength has conventionally been obtained from quaternary and quinternary compound semiconductors [13], [14], [15] and [16], and recently, Ge and Ge-embedded group-IV alloys have shown strong possibilities for providing near-1550-nm optical source [6], [7],[8], [9], [10], [17], [18] and [19]. Thus, now it is advantageous to start considering possible wavelengths for operating an integrated optical system from 1550 nm even when it comes to the on-chip level optical interconnect, which is the reason that this wavelength has been used in the present simulations.

Specified refractive indices of materials at this wavelength were used accordingly. For higher accuracy in the physical parameters obtained by approximate equations taking this specific wavelength and Ge fractions into account, several literature sources were used collectively with self-consistent linkages. For SiO_2, a refractive index (n_{SiO2}) of 1.528 from one of the most recently measured values at 1550 nm was consistently used throughout the simulations [20]. Refractive indices of $Si_{1-x}Ge_x$($n_{Si1-xGex}$) for different values of x were obtained from a second-order polynomial equation for approximation[21], with a set of modified coefficients:

$$n(x) \cong n(x=0)+ax+bx^2 \quad (1)$$

where $n(x = 0)$ is the refractive index of Si (n_{Si}) at 1550 nm, which is 3.477 from a recent study [22]. In order to obtain the remaining unknown coefficients, $x = 1$ and$x = 0.5$ were fed into Eq. (1) to generate 2 first-order simultaneous equations. The refractive index of Ge (n_{Ge}) at 1550 nm, $n(x = 1)$ on the right-hand side of Eq. (1), was 4.275 in establishing the first equation [23]. For the other equation with $x = 0.5$($n_{Si0.5Ge0.5}$), the linear relation between relative permittivity of $Si_{1-x}Ge_x$($\varepsilon_{r,Si1-xGex}$) and x was used [21].

$$\varepsilon_{r,Si_{1-x}Ge_x}(x) = \left[n_{Si_{1-x}Ge_x}(x)\right]^2 = n_{Si}^2 + \left(n_{Ge}^2 - n_{Si}^2\right)x, \quad (2)$$

where $n_{Si} = 3.477$ and $n_{Ge} = 4.275$ were used recursively, and $n_{Si1-xGex}$ was extracted to be the square root of the rightmost term for each x . Inserting the$n_{Si0.5Ge0.5}$ obtained from Eq. (2) back into Eq. (1) for $x = 0.5$ completes the simultaneous equations and results in $a = 1.194$ and $b = -0.396$. During the calculations, the extinction coefficient (k) of $Si_{1-x}Ge_x$ was ignored since the imaginary part of refractive index had little impact on optical confinement and speed. However, the effects of k on optical performance are investigated in the later part of next chapter. The extracted refractive indices (n) are listed in Table 1 and have been reflected throughout the results in the following sections.

Table 1. Calculated refractive indices of $Si_{1-x}Ge_x$ at 1550 nm.

x	$n_{Si1-xGex}$
0	3.477
0.2	3.67
0.4	3.891
0.6	4.051
0.8	4.179
1	4.275

3. SIMULATION RESULTS

3.1. Cohesiveness of optical profile in $Si_{1-x}Ge_x$ waveguide

Fig. 3(a–f) shows the optical mode profile inside the waveguide structure with various Ge fraction x from 0 to 1 in steps of 0.2. All the figures were drawn with the same scale based on colour. Here, the movement of the peak location and the change in optical intensity with Ge fraction are indicated graphically. Consideration of a double-slab structured waveguide consisting of an upper $Si_{1-x}Ge_x$ slab and a lower Si slab provides a way of confirming that the optical mode confined to Si from the whole domain moves further towards Ge slab centre. When the waveguide is composed of Si only, the optical mode peak is located at the centre of the Si waveguide (Fig. 3(a)). As can be confirmed by Fig. 2, the centres of the entire domain, including both the waveguide and cladding silica (line A), and the Si waveguide (line D), are not matched but separated by 125 nm. Although the photons are incident uniformly over the entire domain, the peak of the confined optical mode is located at the Si waveguide centre (line D) at a definite distance from the domain centre (line A) as shown in Fig. 3(a). As Ge fraction increases, the optical mode peak moves from the Si waveguide centre (line D) towards the centre of the upper slab containing Ge (line E), which supports that introducing Ge provides more efficient optical confinement.

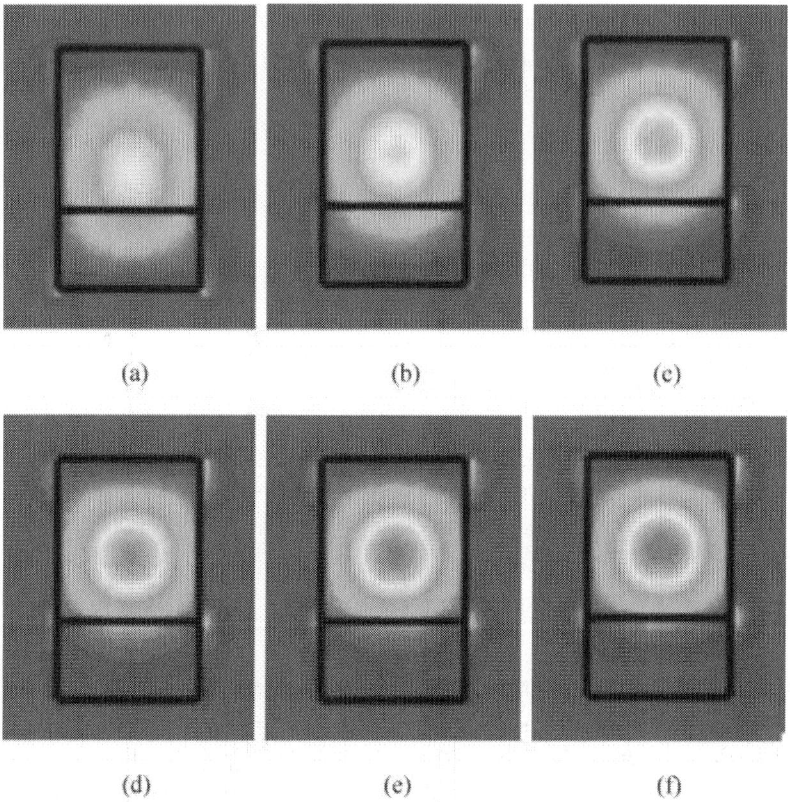

Figure 3. 2-D spatial mode profiles of $Si_{1-x}Ge_x$ waveguide with identical colour scale. (a) $x = 0$, (b) $x = 0.2$, (c)$x = 0.4$, (d) $x = 0.6$, (e) $x = 0.8$, (f) $x = 1$.

Fig. 4(a) shows the optical intensity along the vertical cutline passing through the horizontal centre of the waveguide structure. It is more clearly proven by this figure, in conjunction with Fig. 3(b–f), that the intensity peak increases, and its location moves from line D to line E as x increase. The horizontal axis is scaled by relative location with starting point of the cutline as zero, and the interval between 200 nm and 950 nm is the region of interest where the optical profile is analyzed in the height direction over the waveguide. The boundary of each region and useful reference points are denoted in accordance with Fig. 2. Fig. 4(b) traces the optical intensity peak in greater detail as a function of x, which demonstrates its movement from the centre of the Si waveguide (line D) to that of the $Si_{1-x}Ge_x$ slab (line E). Here, the vertical axis is scaled by relative location starting from one end of cutline and reading information with reference to the centre of entire waveguide (located at 575 nm) and that of the $Si_{1-x}Ge_x$ slab to provide a more tangible understanding. The difference between n_{Ge} and n_{Si} is not as large as that between n_{Si} and n_{SiO2}, so the optical profile is not perfectly confined in the $Si_{1-x}Ge_x$ slab but leaves a fringing distribution behind in the bottom Si. For this reason, the location of the intensity peak does not reach the centre of $Si_{1-x}Ge_x$ slab (located at 450 nm). In any cases, the optical intensity in the silica cladding is negligibly small but a significant amount of light intensity extends into the Si even when $x = 1$, which is confirmed by Fig. 4(a).

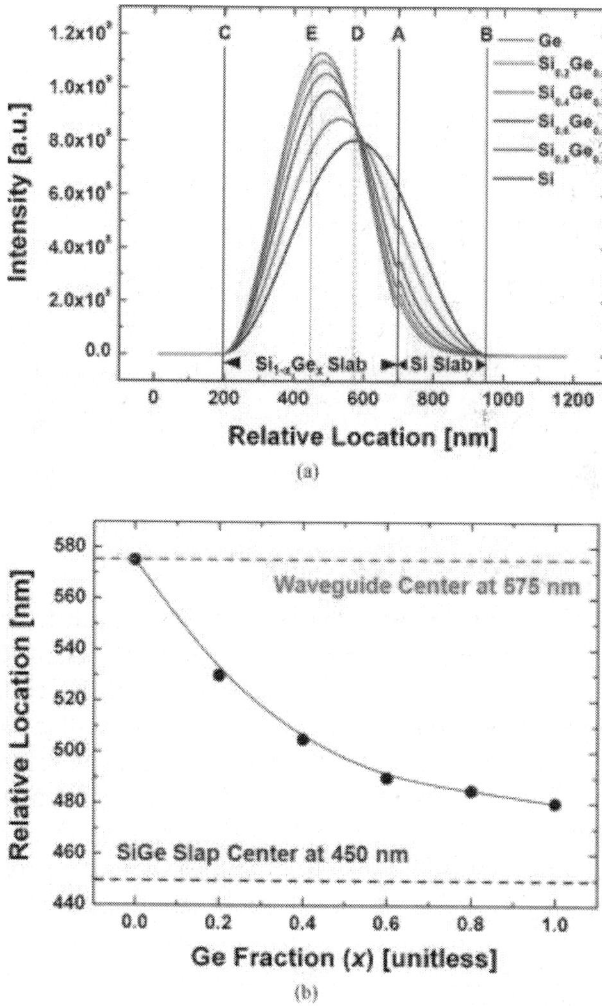

Figure 4. 1-D analysis along the cutline across the centre of $Si_{1-x}Ge_x$ waveguide. (a) Optical intensity profiles. (b) Tracing of intensity peak as a function of x.

3.2. Optical confinement and transmission speed

Fig. 5(a–f) are 3-dimensional (3-D) plots of optical intensity over the waveguide plane as x increases from 0 to 1 in steps of 0.2. Each datum was extracted from a point at the intersection of horizontal and vertical grid lines separated by 5 nm over the 450 nm (width) × 500 nm (height) of the area encompassed by the waveguide cross-section. In analysing the optical confinements, the Si slab was removed from the bottom to investigate the dependence of the mode profile on a single material with no optical leakage path except the circumference cladding. Thus, 450 nm × 500 nm is the area intrinsically defined by $Si_{1-x}Ge_x$. As could be predicted by 1-D analysis along the vertical cutline, the optical intensity profile takes on a more concentrative shape, and its peak value becomes higher as x increases over the 2-D domain. 3-D surface mappings in Fig. 5(a) ($x = 0$) through (f) ($x = 1$) strongly supports that the fringing optical components either on the boundary plane or at the corners are noticeably suppressed. Fig. 6 depicts the maximum optical intensity and confinement factor as a function of x. The intensity for each transverse optical mode in an individual 5 nm × 5 nm mesh was computed using the components of the optical field obtained by solving the vector Helmholtz equation:

$$I_m(p,q) = \left(\left[\vec{E}_{m} \times \vec{H}_{m}^{\,*} \right] + \left[\vec{E}_{m}^{\,*} \times \vec{H}_{m} \right] \right) \cdot \hat{z} \tag{3}$$

$$\int I_m(p,q)\, dS = 1 \tag{4}$$

In Eq. (3), the subscript m denotes intensity from a mesh, $\hat{z}$ is the unit vector in the z-direction (propagating direction) which is perpendicular to the waveguide cross-section, and (p, q) is any point on the grid plane and can be understood as (x, y) in Cartesian coordinates (the variable names were replaced to avoid confusion with the Ge fraction, x). Eq. (4) implies a unique manipulation of Im that the simulation runs for convenience in obtaining the confinement factor of an optical structure. The total intensity calculated by areal integration at 1550-nm wavelength (photon energy = 0.8 eV) is normalized by the entire simulated area, which was 1 μm (width) × 3 μm (height) containing both the 450 nm × 500 nm $Si_{1-x}Ge_x$ waveguides at the core and the SiO_2 cladding around it in all directions. On this basis, the maximum values of Im at different Ge fractions were extracted and plotted on the left vertical axis in Fig. 6. Since the overall optical intensity is already normalized by the total area, the local areal integration over the waveguide plane would readily result in a confinement factor which is defined as the ratio optical intensity of confinement to that of incidence[24]. The extracted confinement factors are plotted as a function of *x*, on the right vertical axis in Fig. 6.

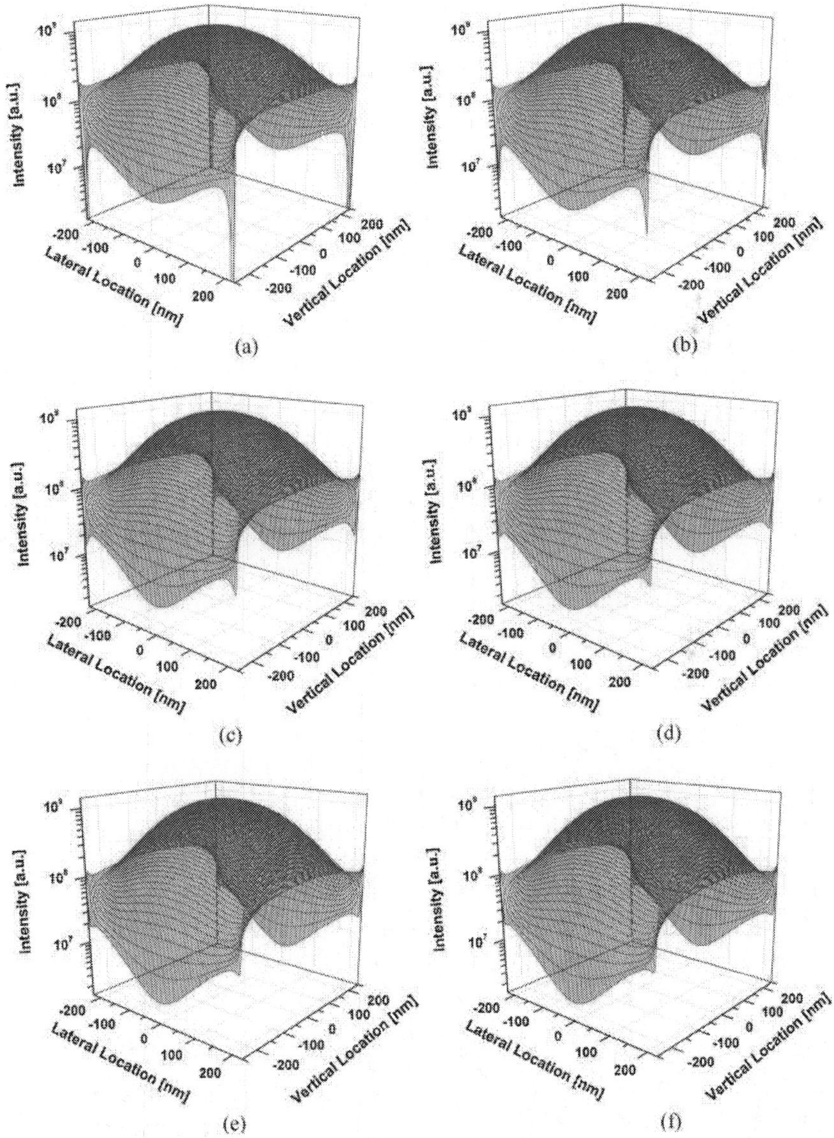

Figure 5. 3-D mapping of optical intensity over the $Si_{1-x}Ge_x$ waveguide plane. (a) $x = 0$, (b) $x = 0.2$, (c) $x = 0.4$, (d) $x = 0.6$, (e) $x = 0.8$, (f) $x = 1$.

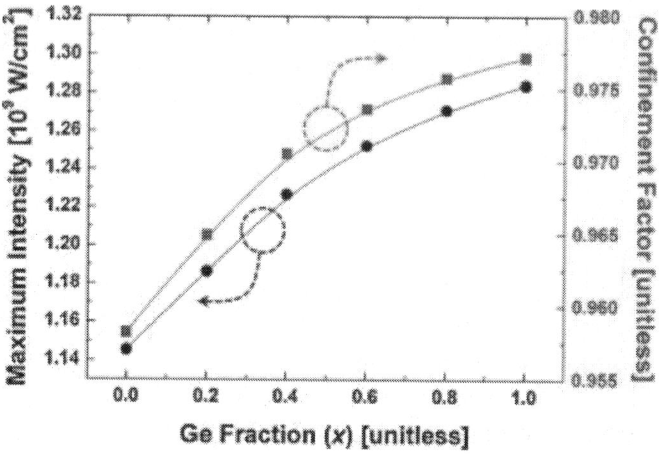

Figure 6. Maximum intensity and confinement factor of $Si_{1-x}Ge_x$ waveguide as a function of *x*.

The value of $n_{Si1-xGex}$ affects maximum bit rate ($BT_{,max}$), which is another requirement for a high-performance waveguide, when it is accommodating multi-mode light signals. In order to calculate $BT_{,max}$ as a function of *x*, the intermodal dispersion is first evaluated by[2]:

$$\delta t = t_{\max} - t_{\min} = \frac{L}{c}\frac{n^2_{Si_{1-x}Ge_x}}{n_{SiO_2}} - \frac{Ln_{Si_{1-x}Ge_x}}{c}$$
$$= \frac{L}{c}\frac{n^2_{Si_{1-x}Ge_x}}{n_{SiO_2}}\left(\frac{n_{Si_{1-x}Ge_x} - n_{SiO_2}}{n_{Si_{1-x}Ge_x}}\right) = \frac{L}{c}\frac{n^2_{Si_{1-x}Ge_x}}{n_{SiO_2}}\Delta \quad (5)$$

Eq. (5) quantifies the time delay difference between the fastest and slowest modes travelling through a waveguide of length *L*. Here, t_{max} is induced from the rays propagating at the critical angle, *c* is the light speed, and Δ is the relative refractive index. Expressing the quantified intermodal dispersion as time delay per unit length ($\delta t/L$) would be more comprehensive. Since we are more focused on either on-chip or on-multi-chip-module (on-MCM) optical interconnect in this work than on long-haul communication, the distances being managed would be no longer than a few tens of millimetres (mm) as is implied by Table 2 (particularly in the columns with thicker lines)[25]. Thus, $\delta t/L$ with units of s/mm provides us with more easily read physical quantities as does the evaluation of $BT_{,max}$ per mm. Since optical pulses must not to overlap with one another, the maximum pulse broadening must be a maximum of half of the transmission period (T), which is often twice the pulse length. Thus, the $BT_{,max}$ is simply as:

$$B_T = \frac{1}{T} \leq \frac{1}{2\delta t} = B_{T,max} \tag{6}$$

Fig. 7 depicts the time delay as intermodal dispersion (left vertical axis) and $BT_{,max}$(right vertical axis) calculated by Eqs. (5) and (6). Although "maximum" is used in the terminology, $BT_{,max}$ can be safely understood as the worst case value we can expect in reality in the sense that the above mathematical inductions have been made under the assumption that the chip has only one waveguide between two components, the source and the detector, over the total length of waveguide. The number of lines should be increased for optical interconnect in highly integrated on-chip/on-MCM systems that operate in a great number of optical and electronic devices as can be inferred by the rightmost columns in Table 2. Consequently, the expectation value of the waveguide length among components is inevitably shortened, which results in a reduced bit rate (BT). Furthermore, for those highly integrated systems, blocks for optical signal synchronization can be adopted, as can be found analogously in a number of technologies for synchronized timing in modern digital circuits [26],[27] and [28]. For these reasons, in real integrated optoelectronic circuits, it is highly probable that actual BT will significantly exceed $BT_{,max}$ in Fig. 7.

Table 2. Dimension and density of optical interconnect in various applications.

	Internet/ WAN	LAN	Rack-to-rack	Card-to-card	On-card	On-MCM	On-chip
Distance	Several kilometer	10–2000 m	30 + m	1 m	0.1–0.3 m	5–100 mm	0.1–10 mm
Number of lines	1	1–10	~100	100–1000	~1000	~10,000	~100,000
Use of optics	Since the 1980s	Since the late 1990s	Now	2010+	2010–2015	Probably after 2015	Later, if ever

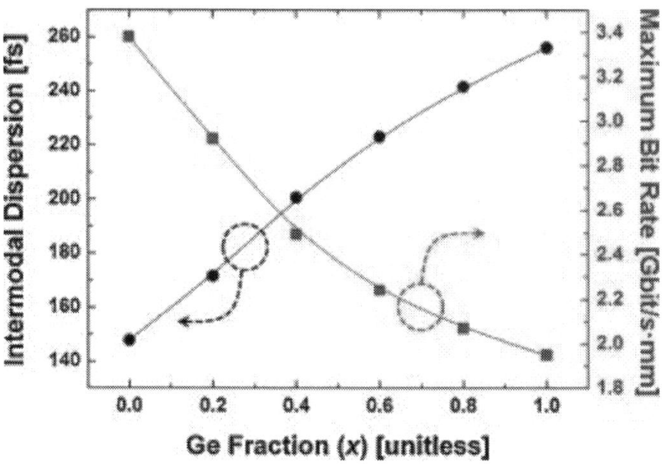

Figure 7. Intermodal dispersion and maximum bit rate of $Si_{1-x}Ge_x$ waveguide as a function of *x*.

3.3. Transmission power compensated by optical confinement

The fraction of the incident power available after propagating a distance through a waveguide material is governed by extinction coefficient, *k*, the imaginary part of complex index of refraction $nc = n - ik$. Although *k* does not have an effect in evaluating performances regarding optical confinement and speed as proceeded in the previous section, it starts to significantly affect the optical loss in passive devices such as waveguide and optical cavities and it is not ignorable any more as Ge fraction increases in $Si_{1-x}Ge_x$ alloy [29]. A recent measurement result shows that *k* of Si ($x = 0$) is 0 above 1120 nm and that of Ge ($x = 1$) is 0.01 at 1550 nm [23]. Although the theoretical maximum *k* value of Ge at 1550 nm (0.8 eV) was predicted to be 6×10^{-3} in a literature by numerical simulation [30], the empirical value, 0.01, will be used for higher credibility in this work. A simulated absorption coefficient (α) of $Si_{0.6}Ge_{0.4}$ with phonon-effect correction is 4 cm^{-1} at photon energy of 0.8 eV [31], which is converted to $k = 4.93 \times 10^{-5}$ from the relation between α and *k* [32]. Due to substantial lack of both experimental and simulation data in the previous literatures, the prediction of *k* values has been accomplished by a mathematical modelling of a closed-form function of Ge fraction at 1550-nm wavelength in this work. *k* of $Si_{1-x}Ge_x$ was modelled as a function of photon energy (*E*), and eventually, that of alloy composition, x [30]. A further modelling can begin with a function in the form of product of a linear function of x and an exponential function to the power of linear function of x.

$$k(x)=(p+qx)\cdot N(s+tx)=(p+qx)\cdot Ns\cdot x(Nt)=(a+bx)\cdot Mx \quad (7)$$

Although there are a number of unknown coefficients in the original function [30], they are lumped into only three coefficients, a, b, and M in Eq. (7), which would have a non-trivial unique solution since $k(x = 0)$, $k(x = 0.4)$, and $k(x = 1)$ are already known from the experiments and physical simulations above.

$$k(0)=(a+b\cdot 0)\cdot M^0=a=0 \quad (8)$$

$$k(0.4)=(a+b\cdot 0.4)\cdot M0.4=0.4b\cdot M0.4=4.93\times 10^{-5} \quad (9)$$

$$k(1)=(a+b\cdot 1)\cdot M^1=bM=0.01 \quad (10)$$

$a = 0$ is obtained immediately by Eq. (8) and solving simultaneous equations of Eqs.(9) and (10) results in $b = 6.58 \times 10^{-6}$ and $M = 1520.62$. The k function is finally obtained as below.

$$k(x)=(6.58\times 10^{-6})x\cdot x(1520.62) \quad (11)$$

Thus, the unknown k values for $x = 0.2$, 0.6, and 0.8 are calculated to be$k(x = 0.2) = 5.69 \times 10^{-6}$, $k(x = 0.6) = 3.20 \times 10^{-4}$,and $k(x = 0.8) = 1.85 \times 10^{-3}$, respectively, from the modelling function in Eq. (11). k of $Si_{1-x}Ge_x$ increases with $x$very rapidly as a result of combining linear and exponential functions of x as confirmed by the semi-empirical modelling result and graphical depiction in Fig. 8.

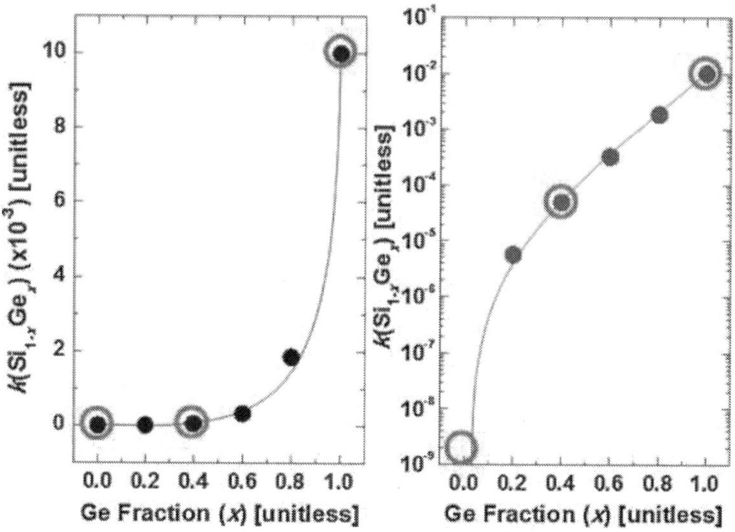

Figure 8. Semi-empirically modelled extinction coefficients (k) of $Si_{1-x}Ge_x$ as a function of x in the linear scale (left) and the logarithmic scale (right). The blank circles indicate the values obtained from experiments and physical simulations [23] and [29].

In order to investigate how effectively higher optical confinement compensates the optical loss, an evaluating function, compensated transmission power (CTP) of SiGe waveguide, can be defined as the product of confinement factor and transmitted power relative to the input after propagating a unit length as formulated in Eq. (12).

$$\text{Compensated transmission power (CTP)} \\ = 10\log\left[\text{Optical confinement} \times \frac{P(x)}{P(0)}\right] [dB] \tag{12}$$

Fig. 9 depicts CTP of SiGe waveguide as a function of Ge fraction with a unit length of 100 μm. As shown in the figure, a Ge fraction up to 72% can be accepted if 3-dB cut-off (half CTP) fraction is considered for the material design. Although it is found from the drastic decrease of CTP that severe optical loss in the SiGe waveguide at a high Ge fraction is not fully compensated by its higher optical confinement, a permissible range of incorporating Ge can be established, which is much wider than that for an optical cavity, as can be shown in the following section.

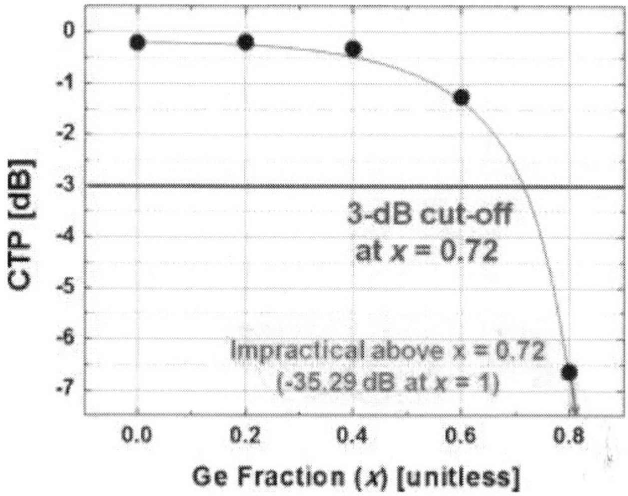

Figure 9. Compensated transmission power (CTP) of SiGe waveguide as a function of Ge fraction.

3.4. Performance of SiGe ring resonator under material loss

Optical resonator is one of the key components comprising the optical interconnect of which cavity performances largely depend on material loss. The effect of *k* varying with Ge faction on quality of $Si_{1-x}Ge_x$ all-pass microring resonator is quantitatively investigated at 1550-nm wavelength and the modelled function in the previous section will be constantly utilized. Among several components making up the optical interconnect system, waveguide-coupled ring resonator is a very simple passive component of which performances as an optical cavity is largely dependent on material loss, and more fundamentally, *k* value of the base material. Fig. 8 shows the schematic of all-pass filter (APF) or notch filter which is the simplest form of a ring resonator, where one output of a directional coupler is fed back into its input. In Fig. 10, *s* and *c* are self- and cross-coupling coefficient, respectively, and s^2 and c^2 are the power splitting ratios of the coupler, and they are assumed to satisfy $s^2 + c^2 = 1$ when no coupling loss is considered [33]. In the following simulation results for evaluating $Si_{1-x}Ge_x$ APF ring resonator, the power splitting between waveguide and resonator is assumed to be 50:50 so that $s = 0.71$ is consistently used. a is the single-pass amplitude transmission, including both propagation loss in the ring and loss in the couplers. It relates to the power attenuation coefficient α [1/cm] as $a^2 = \exp(-\alpha L)$ where L is the round trip length. Absorption coefficients calculated by extinction coefficients from modelling will be used for a values assuming that power attenuation is mainly translated from material loss rather than scattering in the narrow ring structure. For evaluating $Si_{1-x}Ge_x$ APF ring resonator as a function of Ge fraction, x, full width at half maximum (FWHM) and quality factor (Q-factor) will be investigated as two primary characteristic parameters of concern. FWHM and Q-factor for an APF ring resonator are expressed as follows in Eqs. (13) and (14) [33]:

$$\mathrm{FWHM} = \frac{(1-ra)\lambda_{\mathrm{res}}^2}{\pi n_{\mathrm{group}} L\sqrt{ra}} \tag{13}$$

$$\text{Q-factor} = \frac{\lambda_{\mathrm{res}}}{\mathrm{FWHM}} \tag{14}$$

Here, λ_{res} is the wavelength of the light in the ring on resonance and n_{group} is the group index.

$$\lambda_{\mathrm{res}} = \frac{n_{\mathrm{eff}} L}{m}, \quad m = 1, 2, 3, \ldots \tag{15}$$

$$n_{\mathrm{group}} = n_{\mathrm{eff}} - \lambda_0 \frac{dn_{\mathrm{eff}}}{d\lambda} \tag{16}$$

Here, λ_0 is the wavelength of input light in vacuum, 1550 nm. In Eqs. (15) and (16), n_{eff}indicates the effective index of a guiding material determined in the relation with the cladding material, which is formulated as follows [34]:

$$n_{\mathrm{eff}}^2 = n_g^2 b + n_c^2(1-b) \tag{17}$$

ng and nc are the refractive indices of guiding and cladding materials at a given light wavelength, respectively, and b is the confinement factor of the guiding material. $ng$and b values have been obtained in the previous section and can be brought fromTable 1 and Fig. 6. The cladding material is SiO_2 as demonstrated in the simulations and nc is known to be 1.444 at λ_0 = 1550 nm [23]. Optical parameters used for extracting the FWHM and Q-factor of SiGe APF ring resonators with different Ge fractions have been calculated by Eqs. (16) and (17) and the relation $a^2 = \exp(-\alpha L)$. The outer diameter (D) of ring resonator was assumed to be 10 μm so that L becomes$\pi D = 10\pi$ μm. In a previous experimental result, it turned out that a microdisk resonator with $D = 10$ μm had an effective diameter (D_{eff}) of 9.7 μm [19], and$L_{eff} = 9.7\pi$ μm has been consistently used to obtain a values in numerically evaluating the ring resonator performances in this work. $dn_{eff}/d\lambda$ values in Table 3 have been obtained from the first derivatives of Eq. (17) to λ:

$$\frac{dn_{eff}}{d\lambda} = \frac{n_g}{n_{eff}} b \frac{dn_g}{d\lambda} + \frac{n_c}{n_{eff}} (1-b) \frac{dn_c}{d\lambda} \quad (18)$$

where $dng/d\lambda|x_{=0} = -1.088 \times 10^{-4}$ nm^{-1} (Si), $dng/d\lambda|x_{=1} = -4.510 \times 10^{-4}$ nm^{-1} (Ge), and$dnc/d\lambda = -10^{-5}$ nm^{-1} at $\lambda_0 = 1550$ nm are extracted from the measurement data [23]. For extracting $dn_g/d\lambda$ values with intermediate Ge fractions, linear approximation was conducted. In the previous experiments with 10-μm diameter microdisk resonator, optical confinements in the range of mode number $m = 50$–60 were successful so$m = 60$ has been used as a reasonable quantity for generating FWHM and Q-factor from Eqs. (13), (14) and (15). Fig. 11 depicts the FWMH and Q-factor of $Si_{1-x}Ge_x$ ring resonator as a function of x. FWHM shows a monotonic increase with x, which means that the notch appearing in the optical spectrum gets more diffusive as more Ge is incorporated in the SiGe ring resonator. Drastic increase implies that the attenuation by material loss becomes very severe above $x = 0.6$. However, it is noticeable that Q-factor is improved by heightening the Ge fraction up to about $x = 0.4$ and shows a sharp drop above $x = 0.6$. A permissible amount of Ge incorporation is necessary for SiGe optical cavity as well as SiGe waveguide since enhanced optical confinement is one of the requirements asked by both of them. n and k are the competing factors in determining the passive photonic devices and x should be controlled to be 0.4–0.6 for balancing the factors for the $Si_{1-x}Ge_x$ APF ring resonator studied in this work. The exact value can be differed depending on device types and which aspect of performances a designer would put a higher weight in the device. However, it is assured that Ge fraction needs to be optimized based on prepared design criteria to secure desirable performances of passive SiGe photonic devices.

Figure 10. Schematic of waveguide-coupled all-pass ring resonator.

Table 3. Optical parameters for extracting FWHM and Q-factor of SiGe APF ring resonator.

x	n_{eff}	$dn_{eff}/d\lambda$ (nm^{-1})	n_{group}	A
0	3.41	-1.060×10^{-4}	3.574	1
0.2	3.609	-1.733×10^{-4}	3.878	0.999
0.4	3.835	-2.412×10^{-4}	4.209	0.994
0.6	3.998	-3.088×10^{-4}	4.447	0.961
0.8	4.127	-3.767×10^{-4}	4.711	0.796
1	4.225	-4.446×10^{-4}	4.914	0.081

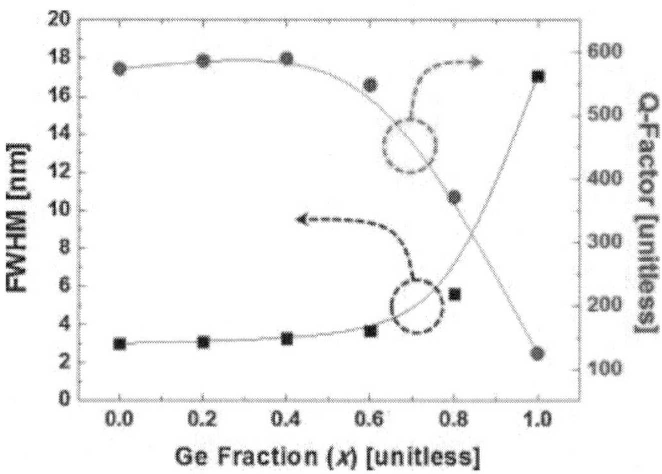

Figure 11. Full width at half maximum (FWMH) and Q-factor of $Si_{1-x}Ge_x$ ring resonator as a function of x.

4. INVESTIGATION OF GERMANIUM INTERFACES IN VARIOUS OPTICAL MATERIAL SYSTEMS

For realizing integrated Si photonic circuits, group-IV alloys or III–V compound materials with recombination efficiency higher than that of Si can be used for optical sources. We examined the interfaces that Ge might form with several candidate materials for achieving the heterogeneous integration on Si platform. Although pure Ge waveguide is not practically preferable due to its excessively high absorption and performance degradation in cavity configuration as confirmed in the previous section, an advantageous feature of Ge-incorporated Si waveguide is found in its better heterogeneous interfacing with some of optical materials of interest. Ge and GaAs have very similar lattice constants, 5.658 Å and 5.653 Å, respectively. The lattice constants of AlGaAs and GeSn monotonically increase as more Al and Sn atoms are introduced in each alloy [35] and [36], and it is clear that $Si_{1-x}Ge_x$ with higher Ge content will demonstrate smaller lattice mismatch with the optical materials. Thus, it would be worthy of having a close look at the interface between $Si_{1-x}Ge_x$ and those materials at an extremity, $x = 1$. By investigating the interfaces, the optical loss by interface roughness of Ge as well as the process viability of Ge in various material systems is predicted. Since n_{Ge} is relatively high, electrically generated light emissions from most of materials are easily introduced to Ge-incorporated waveguide with enhanced optical confinement. Also, a light signal propagating in the $Si_{1-x}Ge_x$ waveguide with a high Ge content can be received by the detectors through butt coupling rather than evanescent coupling for high coupling efficiency [37], [38] and [39]. Spectroscopic analysis of Ge interfaces was performed by transmission electron microscopy (TEM). To investigate the interfaces, TEM cross-section specimens were prepared using conventional cross-sectional method. Each sample was ground and polished down to 40 μm and further ground using a Gatan 656 dimple grinder. Then, it was ion-milled using a Gatan 691 precision ion polishing system (PIPS). A FEI Titan 80–300 environmental TEM was employed for cross-sectional bright-field TEM images at an accelerating voltage of 300 kV. Fig. 12(a–d) demonstrate the analyzed interfaces between Ge/$Al_{0.3}Ga_{0.7}As$, Ge/$Ge_{0.96}Sn_{0.04}$, and Ge/Si, in sequence. In Fig. 12(a), $Al_{0.3}Ga_{0.7}As$ was grown on Ge substrate by metal–organic chemical vapor deposition (MOCVD). The epitaxially grown $Al_{0.3}Ga_{0.7}As$ layer on top of the Ge substrate has very clear and flat interface. In Fig. 12(b–d), $Ge_{0.96}Sn_{0.04}$ on Ge and Ge on Si were grown by reduced-pressure chemical vapor deposition (RPCVD) at low temperatures, 400 °C and 350 °C, respectively, to minimize the migration of stannum (Sn) atoms to the surface and the dislocations between Si and Ge directly grown without SiGe buffer. Ge/$Ge_{0.96}Sn_{0.04}$ interface is indicated as a relatively dark broad line in Fig. 12(b). In Fig. 10(c) Scanning TEM energy-dispersive X-ray spectroscopy (STEM-EDX) mapping of Sn–K peak was performed at the interface of Ge/$Ge_{0.96}Sn_{0.04}$ using a FEI Tecnai G2 F20 TEM at 200 kV. The difference of interface angle is from different sample rotation during STEM image acquisition. Although the image is quite noisy, it still shows a clear and flat interface. The high-resolution TEM (HRTEM) image and EDX mapping image confirm a good Ge/$Ge_{0.96}Sn_{0.04}$ interface with neither Sn precipitation nor phase separation from the GeSn alloy. Also, few threading dislocation was found, owing to the precisely controlled epitaxial growth at a low temperature. Fig. 12(d) shows the epitaxial growth of the Ge layer on the Si substrate clearly from the array of atomic columns of Si and Ge. The value of d-spacing of Si (1 1 1) and Ge (1 1 1) is correspond to the measure lattice fringe value (3.13 Å and 3.26 Å, respectively) from the image which confirms Si/Ge structure. Along the interface, there are island-shaped misfit dislocations due to the strain from lattice mismatch between Si and Ge. Since Ge has much larger lattice constant than that of Si with 4.2% difference, an extra half plane of Si can be seen within the island-shaped area. Overall, the Si/Ge interface was very clean with roughness below only a few atomic layers and the height of a remarkable misfit dislocation (shown in the inset) island was below 1.8 nm. It is confirmed that the crystalline orientations are preserved across the materials and interface roughness is hardly observed under the precisely controlled process conditions.

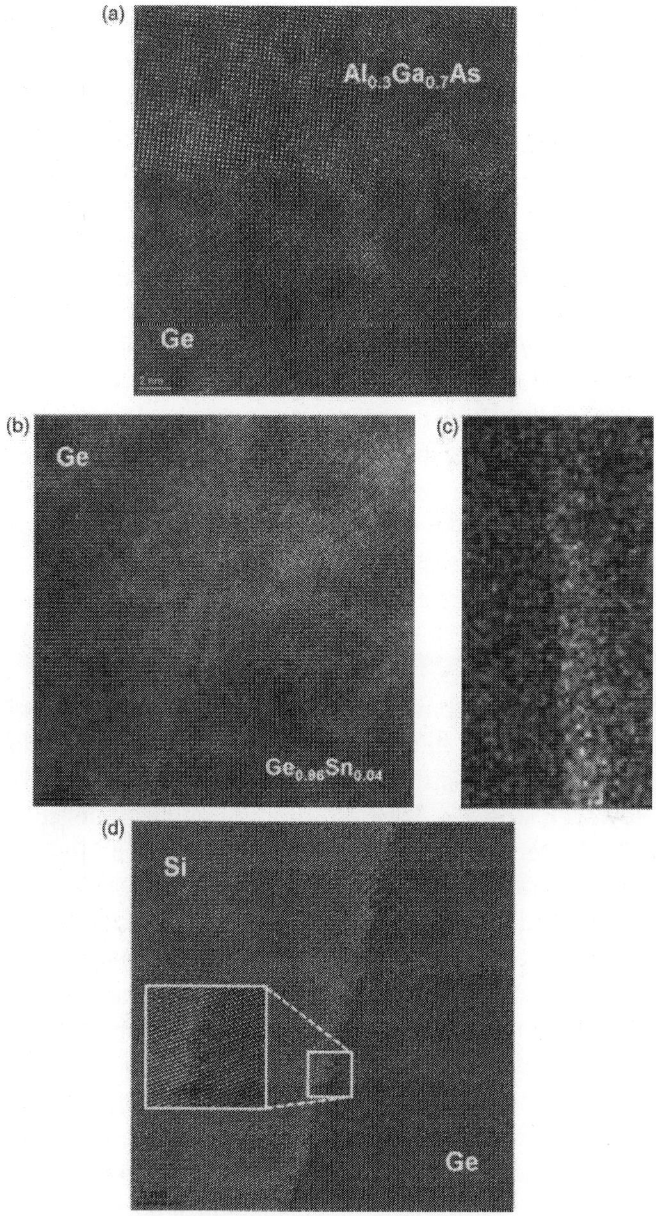

Figure 12. TEM and STEM-EDX images from heterojunction interfaces. (a) Ge and $Al_{0.3}Ga_{0.7}As$. (b) Ge and $Ge_{0.96}Sn_{0.04}$. (c) Sn mapping at the interface of Ge and $Ge_{0.96}Sn_{0.04}$. (d) Ge and Si.

At a given roughness mainly determined by material growth and etching processes, interface scattering loss is reduced as waveguide thickness is increased. However, when it comes to sub-micron dimension, the interface roughness plays a more critical role in scattering loss of planar waveguide [40], [41] and [42]. Although there have been deviations among literatures, a simple model with an exponential correlation function is providing a relatively accurate results for dimensions of the order of the wavelength [42]: for waveguide thickness of 1 μm and RMS interface roughness of 1 nm, interface loss for TE_0 and TM_0 are calculated to be 1.11×10^{-2} dB/mm and 6.30×10^{-3} dB/mm, respectively. The loss is proportional to the square of RMS interface roughness and approximately 100 times larger loss is expected as the waveguide becomes 5 times thinner. Although interface loss should be larger due to increased number of surfaces if the planar waveguide is designed in a 3-D structure, interface losses of the order of 10^{-2} dB/mm would be warranted judging from the microscopic analysis on Ge interfaces.

CONCLUSIONS

Introducing Ge in the conventional Si passive photonic devices will provide better interfacing between Si platform and group-IV alloys or III–V compound semiconductors for monolithically integrated photonic circuits. However, the Ge fraction should be carefully controlled to achieve both better interface and required device performances. The effects of incorporating Ge on performances of SiGe passive devices including waveguide and microring resonator have been closely investigated by optical simulations, mathematical modelling, and the state-of-the-art microscopy technique. Optical confinement and optical loss are the major competing factors in determining the device performances. It has been proven that optical confinement is improved by 2.1% compared with Si waveguide by incorporating Ge and that the half-transmission power compensated by the enhanced confinement factor puts the upper limit (−3 dB cut-off) in Ge fraction at 0.72. Also, it has been demonstrated from the modelling and simulation that reliability indices including FWHM and Q-factor of SiGe microring resonator are degraded above $x = 0.5$. The series of results support that a permissible range of Ge fraction can be set depending on device types and specific requirements. Also, TEM imaging has been performed on Ge boundaries interfacing $Al_{0.3}Ga_{0.7}As$, $Ge_{0.96}Sn_{0.04}$, and pure Si grown by epitaxy, where interface roughness below 2 nm is observed. Through the microscopic analyses, versatile process viability of Ge with Si, group-IV alloys, and III–V compound and interface loss as low as the order of 10^{-2} dB/mm of SiGe waveguide have been confirmed. With optimally incorporated Ge fractions, SiGe based passive devices will play crucial roles in the group-IV photonic integrated circuits.

Competing interests

The authors declare no competing financial interests.

ACKNOWLEDGEMENTS

This work was supported by the Centre for Integrated Smart Sensors funded by the Korean Ministry of Science, ICT and Future Planning as Global Frontier Project (CISS-2012M3A6A6054186). Also, the TEM study was supported as part of the Centre on Nanostructuring for Efficient Energy Conversion (CNEEC), an Energy Frontier Research Centre funded by the US Department of Energy, Office of Science, Office of Basic Energy Sciences under Award Number DE-SC0001060.

REFERENCES

1. Interconnect, International Technology Roadmap for Semiconductors, 2011, 52–62.
2. G.T. Reed, A.P. Knights, Silicon Photonics, John Wiley and Sons, New York, 2005.
3. R.J. Walters, G.I. Bourianoff, H.A. Atwater, Field-effect electroluminescence in silicon nanocrystals, Nature 4 (2005) 143–146.
4. Y. Berence´n, O. Jambois, J.M. Ramı´rez, J.M. Rabled, S. Estrade´, F. Peiro´, C. Domı´nguez, J.A. Rodrı´guez, B. Garrido, Blue-green to near-IR switching electroluminescence from Si-rich silicon oxide/nitride bilayer structures, Optics Letters 36 (2011) 2617– 2619.
5. T. Kessler, C. Hagemann, C. Greibing, T. Legero, U. Sterr, F. Riehle, M.J. Martin, L. Chen, J. Ye, A sub-40-mHz-linewidth laser based on a silicon single-crystal optical cavity, Nature Photonics 6 (2012) 687–692.
6. J. Liu, X. Sun, R. Camacho-Aguilera, L.C. Kimerling, J. Michel, Ge-on-Si laser operating at room temperature, Optics Letters 35 (2010) 679–681.
7. S. Saito, K. Oda, T. Takahama, K. Tani, T. Mine, Germanium fin light-emitting diode, Applied Physics Letters 99 (2011), 241105-1–241105-3.
8. X. Xu, T. Tsuboi, T. Chiba, N. Usami, T. Maruizumi, Y. Shiraki, Silicon-based current-injected light emitting diodes with Ge selfassembled quantum dots embedded in photonic crystal nanocavities, Optics Express 20 (2012) 14714–14721.

9. J. Justice, C. Bower, M. Meitl, M.B. Mooney, M.A. Gubbins, Wafer-scale integration of group III–V lasers on silicon using transfer printing of epitaxial layers, Nature Photonics 6 (2012) 610–614.
10. S. Cho, B.-G. Park, C. Yang, S. Cheung, E. Yoon, T.I. Kamins, S.J.B. Yoo, J.S. Harris, Room-temperature electroluminescence from germanium in an Al0.3Ga0.7As/Ge heterojunction lightemitting diode by G-valley transport, Optics Express 20 (2012) 14921–14927.
11. ATLAS User's Manual, Silvaco International, Santa Clara, 2012.
12. B.G. Streetman, S. Banerjee, Solid State Electronic Devices, Prentice Hall, New Jersey, 2005.
13. V. Lordi, H.B. Yuen, S.R. Bank, J.S. Harris, Quantum-confined Stark effect of GaInNAs(Sb) quantum wells at 1300–1600 nm, Applied Physics Letters 85 (2004) 902–904.
14. M.A. Wistey, S.R. Bank, H.P. Bae, H.B. Yuen, E.R. Pickett, L.L. Goddard, J.S. Harris, GaInNAsSb/GaAs vertical cavity surface emitting lasers at 1534 nm, Electronics Letters 42 (2006) 282–283.
15. R. Kudrawiec, P. Poloczek, J. Misiewicz, H.P. Bae, T. Sarmiento, S.R. Bank, H.B. Yuen, M.A. Wistey, J.S. Harris Jr., Contactless electroreflectance of GaInNAsSb/GaNAs/GaAs quantum wells emitting at 1.5–1.65 mm: broadening of the fundamental transition, Applied Physics Letters 94 (2009), 031903-1–031903-3.
16. T. Sarmiento, H.P. Bae, T.D. O'Sullivan, J.S. Harris Jr., GaAsbased 1.53 mm GaInNAsSb vertical cavity surface emitting lasers, Electronics Letters 45 (2009) 978–979.
17. H. Lin, R. Chen, W. Lu, Y. Huo, T.I. Kamins, J.S. Harris, Investigation of the direct bandgaps in Ge1xSnx alloys with strain control by photoreflectance spectroscopy, Applied Physics Letters 100 (2012), 102109-1–102109-4.
18. R. Chen, H. Lin, Y. Huo, C. Hitzman, T.I. Kamins, J.S. Harris, Increased photoluminescence of strain-reduced, high-Sn composition Ge1xSnx alloys grown by molecular beam epitaxy, Applied Physics Letters 99 (2011), 181125-1–181125-3.
19. S. Cho, R. Chen, S. Koo, G. Shambat, H. Lin, N. Park, J. Vucˇkovic´, T.I. Kamins, B.-G. Park, J.S. Harris, Fabrication and analysis of epitaxially grown Ge1xSnx microdisk resonator with 20-nm free-spectral range, IEEE Photonics Technology Letters 23 (2011) 1535–1537.
20. G. Ghosh, Dispersion-equation coefficients for the refractive index and birefringence of calcite and quartz crystals, Optics Communications 163 (1999) 95–102.
21. M.E. Levinshtein, S.L. Rumyantsev, M.S. Shur, Properties of Advanced Semiconductor Materials: GaN, AlN, InN, BN, SiC, SiGe, John Wiley and Sons, New York, 2001.

22. M. Bass, C. DeCusatis,J. Enoch, V. Lakshminarayanan, G. Li, C. MacDonald, V. Mahajan, E.V. Stryland, Handbook of Optics, 3rd edition. Volume IV: Optical Properties of Materials, Nonlinear Optics, Quantum Optics, McGraw-Hill, New York, 2009.
23. Refractive Index Database (available at http://refractiveindex.- info/).
24. B.E.A. Saleh, M.C. Teich, Fundamentals of Photonics, John Wiley and Sons, New York, 2007.
25. A selected clip from a technology roadmap (available at http:// www.heise.de/ct/meldung/Optische-Verbindung-fuer-CMOSBauelemente-1145306.html).
26. E.G. Friedman, Clock distribution networks in synchronous digital integrated circuits, Proceedings of the IEEE 89 (2001) 665–692.
27. R. Dobkin, R. Ginosar, C.P. Sotiriou, High rate data synchronization in GALS SoCs, IEEE Transactions on Very Large Scale Integration Systems 14 (2006) 1063–1074.
28. T.-Y. Kim, T. Kim, Post silicon management of on-package variation induced 3D clock skew, Journal of Semiconductor Technology and Science 12 (2012) 139–149.
29. R. Braunstein, A.R. Moore, F. Herman, Intrinsic optical absorption in germanium–silicon alloys, Physical Review 109 (1958) 695–710.
30. J. Humlı´c´ek, A. Ro¨seler, T. Zettler, M.G. Kekoua, Infrared refractive index of germanium–silicon alloy crystals, Applied Optics 31 (1992) 90–94.
31. J.L. Polleux, C. Rumelhard, Optical absorption coefficient determination and physical modelling of strained SiGe/Si photodetectors, in: Proceedings of 2000 8th IEEE International Symposium on High Performance Electron Devices for Microwave and Optoelectronic Applications, 2000, pp. 167–172.
32. J.I. Pankove, Optical Processes in Semiconductors, Dover Publications, Inc., New York, 1975.
33. W. Bogaerts, P.D. Heyn, T.V. Vaerenbergh, K.D. Vos, S.K. Selvaraja, T. Claes, P. Dumon, P. Bienstman, D.V. Thourhout, R. Baets, Silicon microring resonators, Laser and Photonics Reviews 6 (2012) 47–73.
34. J. Vucˇkovic´, Class Note from Introduction to Nanophotonics and Nanostructures, Stanford University, 2010 (chapter 2).
35. Basic Parameters of AlxGa1xAs at 300 K (available at http:// www.ioffe.ru/SVA/NSM/Semicond/AlGaAs/basic.html).
36. H.P.L. de Guevara, A.G. Rodrı´guez, H. Navarro-Contreras, M.A. Vidal, Ge1xSnx alloys pseudomorphically grown on Ge(0 0 1), Applied Physics Letters 83 (2003) 4942–4944.
37. D. Feng, S. Liao, P. Dong, N.-N. Feng, H. Liang, D. Zheng, C.-C. Kung, J. Fong, R. Shafiiha, J. Cunningham, A.V. Krishnamoorthy, M. Asghari, High-speed Ge photodetector monolithically integrated with large cross-section silicon-on-insulator waveguide, Applied Physics Letters 95 (2009), 261105-1– 261105-3.
38. S. Feng, Y. Geng, K.M. Lau, A.W. Poon, Epitaxial III–V-onsilicon waveguide butt-coupled photodetectors, Optics Letters 37 (2012) 4035–4037. S.

39. S. Cho, H. Kim, M.-C. Sun, B.-G. Park, J.S. Harris Jr., Process considerations for 80-GHz high-performance p–i–n silicon photodetector for optical interconnect, Journal of Semiconductor Technology and Science 12 (2012) 370–376.
40. P.K. Tien, Light waves in thin films and integrated optics, Applied Optics 10 (1971) 2395–2413.
41. J.P.R. Lacey, F.P. Payne, Radiation loss from planar waveguides with random wall imperfections, IEE Proceedings 137 (1990) 282–288.
42. F. Ladouceur, J.D. Love, T.J. Senden, Effect of side wall roughness in buried channel waveguides, IEE Proceedings Optoelectronics 141 (1994) 242–248.